W9-CUK-307

CRYSTAL STRUCTURE
AND CHEMICAL BONDING IN
INORGANIC CHEMISTRY

CRYSTAL STRUCTURE AND CHEMICAL BONDING IN INORGANIC CHEMISTRY

Proceedings of an International Symposium,
Wageningen, 21-22 February 1974, The Netherlands

Edited by

C.J.M. Rooymans
Philips Research Laboratories
Eindhoven, The Netherlands

A. Rabenau
Max-Planck-Institut
für Festkörperforschung
Stuttgart, F.R.G.

1975

NORTH-HOLLAND PUBLISHING COMPANY - AMSTERDAM - OXFORD
AMERICAN ELSEVIER PUBLISHING COMPANY, INC. - NEW YORK

√62311980

CHEMISTRY

© NORTH-HOLLAND PUBLISHING COMPANY − 1975

All Rights Reserved. No part of this publication may be reproduced, stored in a retrieval system, or transmitted, in any form or by any means, electronic, mechanical, photocopying, recording or otherwise, without the prior permission of the copyright owner.

North-Holland ISBN: 0 7204 0340 5
American Elsevier ISBN: 0 444 10961 7

Published by:
NORTH-HOLLAND PUBLISHING COMPANY − AMSTERDAM
NORTH-HOLLAND PUBLISHING COMPANY, LTD. − OXFORD

Distributors for the U.S.A. and Canada:
American Elsevier Publishing Company, Inc.
52 Vanderbilt Avenue
New York, N.Y. 10017

PRINTED IN THE NETHERLANDS

Q D
921
C77
CHEM

EDITORIAL

    The 80ieth birthday of Prof.Dr. A.E. van Arkel stimulated
the Division for Inorganic and Physical Chemistry of the Royal
Duch Chemical Society (K.N.C.V.) and the Division for Solid
State Chemistry of the German Chemical Society (G.D.Ch.) to
organize a symposium on the various aspects of the chemical bond,
the crystal structure and the properties of inorganic compounds,
a theme which has always been central in the work of A.E. van
Arkel and still is very important in solid state research.
Realizing that complete coverage of this wide and challenging
field was inconceivable, the idea was to have competent and
representative scientists giving up-to-date views on both the
experimental and theoretical aspects of the main areas.
    The resonance of the conference, which draw a large
attention and the many requests from participants and from
outside brought the editors to collect the papers in the form
given. The camera-ready technique has been chosen with the
objective of a price which makes these papers accessible to a
large group of people, including the graduate student level.
    After some introductory remarks, his contemporary
Prof. Klemm gives a historical description of the development
of the electrostatic aspects of the chemical bond. Bongers
subsequently discusses the magnetic properties in relation to
structure and bonding in some well-chosen group of compounds.
Simon demonstrates the use of sophisticated experimental techniques
both in preparation and measurement in order to increase our
insight in the phase relationships and structures of the group
of alkali metal suboxides. Hagenmüller shows on some examples
of vanadium bronzes the application of electrical and magnetic
measurements to deduce the valency state of the cations involved.
Structure, disorder and deviations from stoichiometry in the
solid ionic conductors β -alumina and stabilized zirconia,
a subject of a strong current interest are surveyed by Roth.
The existence of the trigonal-prismatic coordination in several
transition metal chalcogenides is discussed by Haas against the
background of the nature of the chemical bonds involved.
A rather full review on the capabilities of the MAPLE-concept
(Madelung Part of the Lattice Energy) for the preference of
certain structure types is given by Hoppe.

The metallic bond is discussed by Miedema and coworkers, who deduce from first principles simple and very widely applicable rules for the behaviour of metals in binary systems, in excellent agreement with the experimental data. A quantum-mechanical ab-initio treatment of the chemical bond is discussed by Nieuwpoort. Janssen, finally, elucidates the importance of triple interactions within the concept of an ionic model which makes it possible to explain the preference for certain crystal structures and physical properties in binary compounds.

The Editors do hope that this publication will find its place not only among the specialists in the area of solid state physics and chemistry, but certainly as well amongst the students in this still intriguing field of research where relations between physical properties, chemical bond and crystal structure will remain central for many years to come.

CONTENTS

Crystal Structure and Chemical Bonding in Inorganic Chemistry
Eds. C.J.M. Rooymans and A. Rabenau
© 1975, North-Holland Publishing Company, The Netherlands

FOREWORD

PROF. VAN ARKEL AS AN INDUSTRIAL RESEARCH SCIENTIST

H.J. Vink
Philips Research Laboratories
Eindhoven, The Netherlands

A large part of the technical achievements of the last four decades is based upon a special kind of multidisciplinarian research,
that has now become to be designated by the name of Materials
Science. This extended and varied field of science tries to
understand and sometimes to predict the relation between various
combinations of physical and chemical properties of a material
and its chemical composition. That is the relation between those
properties, and the concentration and the valency of the various
component atoms on the available crystallographic sites, taking
into account the crystallographic structure(s) and their
texture. Two main aspects, therefore, can be distinguished in
this branche of science, although they are strongly correlated
and in fact inseparable. One is of a more physical character,
whereas the other is of a more chemical nature.
Professor Van Arkel was one of the first scientists to realize
the necessity of founding the study of inorganic and also organic
compounds on a multidisciplinarian basis. Professor Klemm in
his Introduction to the Work of Professor Van Arkel, with which
this collection of conference papers opens, stresses this fact
most emphatically.
It is striking also to learn that Van Arkel, who started his
career as a colloid chemist with Professor Kruyt and Professor
Bungenberg de Jong, turned his attention to the study of the
relation between physical properties and chemical composition,
after he entered the Philips Research Laboratories as a
research chemist. The electronics industry was in fact one of
the first industries to recognize the importance of fundamental
scientific research within the walls of an industrial
enterprise.
In the sequence of papers first appearing in the 1929 volume of
the "Chemisch Weekblad" under the title "The Chemical Bond as an

Electrostatic Phenomenon", Van Arkel and De Boer in fact did
study the relation between various types of physical properties
and the chemical composition of series of chemical compounds. In
these and related studies the foundation was laid by Van Arkel
and De Boer - later to be continued by Verwey and his coworkers -
of a way of thinking that in the course of the following four
decades proved to be very fruitful in the discovery and
understanding of so many important electric, magnetic and
fluorescent materials and also in developing a comprehensive
view of broad areas that is now being denoted under the general
term of defect chemistry.

In the study of many phenomena in semiconductors and phosphors
it became apparent that in order to really understand the
relation between physical properties and the presence (often in
exceedingly small concentrations) of specific elements, the
matrix material had to be of an extreme purity. The importance
of this was realized by Van Arkel at an early stage, as is shown
by his attention to the way of preparing the materials to be
studied.

His paper of 1925 - with De Boer - on the preparation of very
pure Ti-, Zr-, Hf- and Th-metal, was the start of a long series
of studies (by De Boer and Fast) on the preparation of pure
metals.

This method of preparing metals by decomposition of volatile
compounds, nowadays called chemical vapour deposition, is
widely used presently in the fabrication of transistors and
integrated circuits for the deposition of epitaxial (n- or
p-doped) Si-layers.

His book "Reine Metalle" published in 1939 showed his
continuous interest in this aspect of materials science. Van
Arkel realized, however, that not only the purity of a material
can be of significance, but also its crystalline nature and,
because of that, the availability of single crystals. His paper,
in 1923, on single crystal tungsten must be mentioned in this
respect. In that publication Van Arkel also started his studies
on the relation between the properties and the texture of a
material. His investigations on recrystallization in W, Al and
Sn are ample proof of Van Arkel's interest in these complicated
matters.

It is understandable that in Van Arkel's way of thinking the
constant use of the theory of phase relations in its widest
aspects is indispensable. In his lectures on physical chemistry,
Professor Van Arkel made it abundantly clear to his pupils, that
one should never neglect this field of general chemistry.
With a slight change in the wording of one of the sentences in
Professor Klemm's paper on the Development of the Electrostatic
Aspects in Chemical Bonding, it can truly be said that the
beginnings of nearly all the problems, which are relevant in
that branch of chemistry called materials science, can be found
in the papers of Van Arkel.
It has been a very good idea to celebrate Van Arkel's 80th
birthday by a conference dealing with the present day aspects
of the chemical bond, the crystal structure and the properties
of inorganic compounds, a theme which was always central in his
work. His influence on the development of concepts of chemical
bonding has been great indeed. He has shown, as we all know,
how, by careful reasoning starting from a few simple arguments,
one may get the whole complex field of inorganic chemistry into
one's grip. Therefore the subject of this celebration conference
has been rightly chosen. Professor Klemm's contribution leaves
no doubt about this important aspect of Van Arkel's activities.
It is for me to throw light on another aspect of his work.
Professor Van Arkel opened up that field of chemistry of which
the present days aspects have been discussed during this
conference at the Philips Research Laboratories.
As a long time member of these same laboratories I am allowed,
I think, to take this opportunity to describe the second aspect
of his work as that of a materials scientist "avant la lettre".

Crystal Structure and Chemical Bonding in Inorganic Chemistry
Eds. C.J.M. Rooymans and A. Rabenau
© 1975, North-Holland Publishing Company, The Netherlands

INTRODUCTION TO THE WORK OF PROF. VAN ARKEL

Wilhelm Klemm, Professor-emeritus

University Münster, 44 Münster, Germany

The Wageningen symposium has been organized to honour
Prof. van Arkel on the occasion of his eightieth birthday. It is
my privilege to give an introductory chapter concerning the
scientific work of the man who is celebrating his jubilee.

I have known Prof. van Arkel for forty five years. In 1929
he and the late Prof. de Boer visited Hannover. Both gentlemen
gave lectures: on problems of recrystallization and on their
famous method of preparing metals like zirconium and hafnium by
decomposition of the iodides on a hot wolfram wire.
This new method was a highlight of preparative inorganic chemistry
at the end of the twenties. Thanks to the generosity of our
colleagues in Eindhoven, a great number of institutes in Germany
had such rods of titanium in their collection and demonstrated
them with pride. I personally received samples of TiO, TiN and
TiC for magnetic measurements.

This meeting in Hannover had a very personal character
- owing to the hospitality of Wilhelm Biltz. I learned, that
Prof. van Arkel began as a colloid chemist having been a pupil
of Prof. Kruyt. I had the privilege to meet this outstanding
scientist a number of times as he was very active in international
scientific organizations. The first publications of van Arkel
were concerned with colloid chemistry; the situation changed,
however, when he was appointed a research chemist at the
Philips Laboratories. These laboratories in Eindhoven were and
are one of the most well known places of inorganic research in
the world. The application of physical methods to inorganic
problems which has become of fundamental importance for the
'renaissance' of inorganic chemistry, was carried out here very
early. The members of the scientific staff had - in addition to
their work on problems of technical interest - the possibility
of doing basic research. Obviously, the investigations on the
recrystallization of metals and new methods for the preparation
of metals were largely determined by technical needs; this work
included the preparation of pure compounds from which these

metals could be made - e.g. the separation of zirconium from
hafnium. In addition mixed crystals and the lattice structure
of several elements and compounds were investigated and adsorption
phenomena were studied.

The realization that many properties of metals are strongly
influenced by impurities may have been the reason why van Arkel
- with the co-operation of many experts from different countries -
in 1939 published the book 'Reine Metalle' which was of great
use to metallurgists and chemists in science and technology.

One year after the meeting in Hannover, in 1930, I met
van Arkel and de Boer again. At that time a large number of
chemists from the Netherlands attended the meetings of the
'Nordwestdeutsche Chemiedozenten' and therefore the Dutch
colleagues invited the German group to have a joint meeting in
Amsterdam. Together with my late wife, who was also a chemist,
I attended this meeting. Here we learnt that van Arkel and
de Boer had written a series of papers in Chem. Weekblad having
the title 'De Chemische Binding als Electrostatisch Verschijnsel';
these papers had been published as a book by Centen (Amsterdam).
We were rather astonished to learn that scientists working in
an industrial laboratory were so much interested in theoretical
problems. Still more we were surprised that the Dutch colleagues
proposed to us to translate the book into German. We were
reluctant because we did not understand a single word of the
Dutch language. But van Arkel and de Boer encouraged us: Dutch
and German are not too different, so we would have no difficulty.
We promised to make an attempt and it was succesful. The German
edition appeared in 1931. The book had a very wide reception in
Germany and promoted the understanding of the electrostatic
theory of chemical bonding to a high degree. One must bear in
mind that the new theory had been developed by physicists and
that the majority of the papers had been published in physical
journals. These papers had to be 'translated' into a language
understood by chemists. That is what van Arkel and de Boer had
done. Together with a survey of the literature, they gave their
own calculations and demonstrated the fields of chemistry in
which the application of the electrostatic theory would be useful,
and they also made clear the limitations of the theory! - A
French translation appeared in 1936.

Later on Prof. van Arkel published a similar book *'Moleculen
en Kristallen'* (van Stockum, 's-Gravenhage). Five editions of
this book have been published between 1941 and 1961; the second
edition has been translated into English. In the next paper
I shall give a survey of the main topics treated in these books.
But before doing so, let me give a short outline of the further
activities of Prof. van Arkel. It was natural that this intense
occupation with the electrostatic theory of chemical bonding
influenced his experimental work. One problem which interested
him very much was the influence of electric dipoles on the
physico-chemical qualities of liquids (cohesion, boiling points,
solubility, etc.). At it was much easier to find suitable
examples among the organic compounds, van Arkel made these
investigations with organic liquids. It is typical of him, that
his interest did not stop at artificial boundaries between
different disciplines.

In 1934 an incisive change in the career of van Arkel took
place; he left Philips and accepted the chair of Inorganic and
Physical Chemistry at the University of Leyden. At this time I
left Hannover and moved to Danzig, which is rather far away from
the Netherlands. So it was natural that we met one another
rather rarely after that. Finally the war cut off our relations
completely. Therefore, I am pleased to say that the warmhearted
article published by E.W. Gorter and F.C. Romeyn in Chemisch
Weekblad 60 (1964), 298-308, provided the basis for my remarks
about van Arkels activity as professor in Leyden.

It is obvious that his great interest in educating young men
in natural science was the reason for van Arkels decision to
trade a very successful career in an industrial research
laboratory for a chair at a university. Of course, it cannot
be my task here to discuss the new organization of chemistry
teaching introduced by van Arkel at Leyden. His interest in
this problem can be recognized by the fact that some articles
published in the years 1934-1936 are concerned with problems
of teaching, and he wrote - together with H.G.S. Snijder - *Leer-
boek der Scheikunde gegrond op atoommodel en periodiek systeem*
(Noordhoff, Groningen).

Concerning van Arkels research work at Leyden, it is astonishing at first glance to see that the number of his publications after 1934 is not very high. One reason for this is the fact that from 1940-1945 the university was closed by the German occupation forces. After this period there was a vivid activity in the laboratories at Leyden university; but most of the scientific results of the investigations carried under the guidance of van Arkel are described in the theses of his students. Although these theses constitute an important part of van Arkels activity they are not to be found in any abstract under his name. May I give some subjects of these theses in order to point out the fields of interest in the institute at that time properties of liquids; dipole-moments; coordination-compounds; spinels and other ternary compounds including their magnetic properties; transition-elements.

The high esteem in which the scientific work of van Arkel was held is demonstrated by the fact that he was several times invited to deliver plenary lectures at International Conferences During the preparation of this article I have read some of these papers as well as some other surveys published by van Arkel, and I was impressed not only by his outstanding knowledge of the facts but also by his talent to deduce general relations from detailed results which, at first glance, seem not to have any connection with each other.

If now at the age of over 80 years Prof. van Arkel looks back on his life as a scientist he can do so with great satisfaction. The results of his research work brought about remarkable progress in Inorganic Chemistry; he has introduced modern aspects of teaching chemistry at the universities of the Netherlands; a great number of pupils see to it that the ideas propagated by him are the basis for their own fruitful research work in science and technology. Thank you very much, Prof. van Arkel!

Crystal Structure and Chemical Bonding in Inorganic Chemistry
Eds. C.J.M. Rooymans and A. Rabenau
© 1975, North-Holland Publishing Company, The Netherlands

DEVELOPMENT OF THE ELECTROSTATIC ASPECTS IN CHEMICAL BONDING

Wilhelm Klemm

Professor-emeritus, University Münster, 44 Münster, Germany

SUMMARY

The purpose of this contribution is to give a survey of the situation of inorganic chemistry in the nineteen thirties and forties, that means in the period during which the principles of the *Electrostatic Theory of Chemical Bonding* were developed.

I. THE ELECTROSTATIC MODEL IN RETROSPECTIVE VIEW

An influence of electric forces on chemical compounds had already been supposed by Berzelius; one hundred and fifty years ago the majority of chemists was convinced that the theory of Berzelius was correct. At that time, however, chemistry was mainly inorganic chemistry. But when more and more organic compounds were prepared and their structure elucidated it became clear that the theory of Berzelius was not applicable to these compounds. The theory of valency was developed by Kekulé and others and it seemed that this new theory was valid not only for organic but also for inorganic compounds. At the end of the eighties the theory of Arrhenius made clear, that in aqueous solutions of salts electrically charged atoms - or groups of atoms - were present, but this discovery had an astonishingly small influence on the theory of chemical bonding. One reason may be, that the interests of chemists at the end of the last century were so much in the direction of organic chemistry that one was convinced that the relatively small number of inorganic compounds were formed according to the same principles; the Periodic Table, established in 1868-70, demonstrated that the 'valence' of the elements was the deciding factor for the composition of compounds. The question of the physical nature of the chemical valence was not raised at this time!

At the end of the last century there was a revival of interest in inorganic chemistry. One reason may have been the amazing discovery of the noble gases. Still more important was the famous work of A. Werner concerning the structure of coordination compounds in which he created a 'stereochemistry' of these substances. The classical rules of valency were by no means

sufficient for the understanding of this group of compounds.
Werner developed the idea that besides the 'main valence' some
sort of 'secondary valences' exists. This theory was accepted
with enthusiasm by the majority of chemists, but in reality
it was full of internal contradictions. I must confess that,
when I was a student, I never came to understand Werner's ideas
fully - nor do I so today! Very important at the end of the last
century was the fact that the influence of physics on chemistry
became increasingly important. Physical chemistry became a
new discipline. This contributed at the same time to a better
understanding of some fields of inorganic chemistry.

In the years 1912-13 two new results of physical research
were of fundamental importance for chemistry:
1. As the result of a development lasting more than two decades,
   the Bohr-theory of the *structure of the atoms* was formulated.
   It stated that:
   a) the atoms are built up of electrically charged elementary
      units, the nuclei, having a positive charge which corres-
      ponds to the atomic number in the periodic system, and
      shells of electrons each of which bears one unit of negative
      charge;
   b) in the microcosmos of the atoms new laws hitherto not known
      in macroscopic systems hold, the Quantum-rules;
   c) the arrangement of the electrons of the atoms in different
      shells provides an understanding of the periodic system.

2. v. Laue, Friedrich and Knipping discovered the *interference
   of X-rays by crystals*. This opened up the way for the eluci-
   dation of the structure of crystals. The very first examples
   investigated demonstrated that the assumption, hithertho
   generally made by chemists, that chemical compounds in all
   physical conditions are composed of molecules is - at least
   for salts - not correct for the solid state.
   These results formed the basis for new thoughts concerning
the chemical bond, the nature of which was an unsolved problem
up to that time. The first theories were developed in 1916 in-
dependently by W. Kossel in Germany and by G.N. Lewis in the USA.
Characteristic for the theory of Lewis was the assumption that
'lone' electrons of different atoms form '*electron pairs* ', thereby

producing a *'covalent bond '*. A system of eight electrons
surrounding an atom, an 'octet', is especially stable. This
theory is of special interest for organic chemistry and will
not be discussed in detail here.

Of much greater importance for the inorganic chemistry
were the ideas developed by Kossel. The most important items
are the following:

1. Salt-like compounds consist of charged atoms which attract
   one another electrostatically ('ionic bond').
2. The formation of ions with the electronic configuration of
   a rare gas is of predominant importance for the composition
   of stable compounds - especially of those formed by elements
   in the neighbourhood of the noble gases.
3. Calculations demonstrate that spheres with opposite charge
   can not only form neutral molecules but also charged complexes, e.g. [AlF_6 $^{3-}$]. Up to a certain number of ligands such
   complexes are energetically stable.
4. Neutral molecules built up of charged particles attract one
   another for electrostatic reasons. In this way double molecules, triple molecules and so on are formed, resulting
   finally in crystal lattices. Solid substances with such
   'coordination lattices' have high melting and boiling points
   and high electric conductivities in the molten state.
5. For the case of a highly charged small cation totally enveloped
   by the anions this mechanism is not possible; instead a
   'molecule lattice' is formed. Such substances have low melting
   and boiling points and are insulators in the molten state.

Table 1 shows that the electrical conductivity follows these
general rules.

Table 1

Molar electrical conductivity (Ω^{-1} cm^2 mole^{-1}) in the molten state at the melting point according to W. Biltz and W. Klemm.

HCl $\sim 10^{-6}$			
LiCl 166	BeCl$_2$ < 0.086	BCl$_3$ 0	CCl$_4$ 0
NaCl 133.5	MgCl$_2$ 28.8	AlCl$_3$ 15.10^{-6}	SiCl$_4$ 0
KCl 103.5	CaCl$_2$ 51.9	ScCl$_3$ 15	TiCl$_4$ 0
RbCl 78.2	SrCl$_2$ 55.7	YCl$_3$ 9.5	ZrCl$_4$?
CsCl 66.7	BaCl$_2$ 64.6	LaCl$_3$ 29.0	HfCl$_4$?
			ThCl$_4$ 16

II. THE BORN-HABER CYCLE AND THE ENTHALPY OF FORMATION

On the basis of the model proposed by Kossel for ionic compounds it is possible to calculate the energy change on formation of molecules and lattices starting from the ions composing the compound in question. Whereas the calculation based on the assumption of rigid spheres is very simple for isolated molecules, the situation is much more complicated for lattices because the mutual interaction of all ions of the lattice must be considered. Madelung made the calculation for some simple lattice structures. In the meantime it has become possible to calculate the 'Madelung-constant' for complicated lattices. This problem will be dealt with in the paper of Hoppe (p. 127).

Ions are of course not rigid spheres; attraction and repulsion forces exist, and at the equilibrium the repulsion forces equal the attraction forces. In this way the compressibility, the thermal expansion and other qualities become under-

standable. According to Born this model requires the introduction
of a factor n-1/n (n ~ 10). The first possibility to compare
these calculations with experimental values is provided
by the *Energy of Sublimation* (see Table 2). The values calculated
in this way were of the right order of magnitude but the
deviations from the measured values were rather large; therefore
no great attention was paid to these calculations.

Table 2

Change of energy for formation of molecules (ΔH_M) and
lattices (ΔH_u) from monovalent ions.

Rigid spheres	Attraction- and repulsion-forces
Molecule: $N_{Lo} \cdot (A_g^+ + B_g^-) = N_{Lo} \cdot AB_g$	
$\Delta H_M = - \dfrac{e^2}{r_g} \cdot N_{Lo}$	$\Delta H_M = - \dfrac{n-1}{n} \cdot \dfrac{e^2}{r_g} \cdot N_{Lo}$
Lattice: $N_{Lo} \cdot (A_g^+ + B_g^-) = N_{Lo} \cdot AB_s$.	
$\Delta H_U = - A \cdot \dfrac{e^2}{r_s} \cdot N_{Lo}$ $r_g = r_s$	$\Delta H_U = - \dfrac{n-1}{n} \cdot A \cdot \dfrac{e^2}{r_s} \cdot N_{Lo}$ $r_g < r_s$
Sublimation: $N_{Lo} \cdot AB_s = N_{Lo} \cdot AB_g$ $\Delta H = - \Delta H_U + \Delta H_M$	
$\Delta H = (A-1) \dfrac{e^2}{r} \cdot N_{Lo}$	$\Delta H = \dfrac{n-1}{n} \cdot N_{Lo} \left[\dfrac{A \cdot e^2}{r_s} - \dfrac{e^2}{r_g} \right]$

Lo = Loschmidt-number; n = exponent of repulsion
A = Madelung-constant.

The attention was much greater when the experimental values of
the *Enthalpy of Formation* of the compounds, starting from the
elements, were compared with the calculated values, making use
of the Born-Haber cycle (fig. 1). The fact that is was possible
to calculate the enthalpy of formation by combining known data
of the elements with the lattice energy made an extraordinary
impression on my generation. We had the feeling that a new period
of chemistry was beginning and today I must say that this feeling
was correct!

Indeed it was at that time a serious disappointment when
it appeared that the method used for the determination

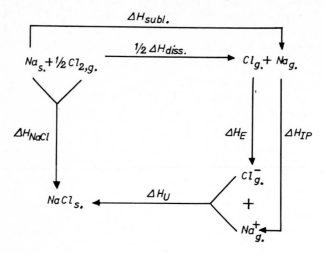

Fig. 1. The Born-Haber cycle.

of the electron affinity ΔH_E was wrong; everything became uncertain again. Fortunately, the electron affinity for one element - iodine - could be determined by an independent method. In this way the situation was made clear at least for the iodides. (Table 3).

Table 3

Determination of the electron affinity of iodine (J.E. Mayer, 1930).

	ΔH
$MI \longrightarrow M^+ + I^-$	H_1
$MI \longrightarrow M + I$	H_2
$M \longrightarrow M^+ + e^-$	H_3
$I + e^- \longrightarrow I^-$	E_I

$$E_I = H_1 - H_2 - H_3$$

In the meantime the values of the electron affinities for monovalent anions - which are exothermic - have been determined very accurately by independent methods. The values for O^{2-}, P^{3-} etc. are strongly endothermic; they must be derived from the Born-Haber cycle.

The possibility to calculate values of the enthalpies of formation by making use of the Born-Haber cycle has sometimes been overemphasized. One must keep in mind that these enthalpies are often the differences of very high numbers and that especially the calculation of the lattice energies may contain large errors. A deviation of a few percent in the lattice energy can lead to a difference of 50 or more calories in the enthalpy of formation! Nevertheless the calculations based on an electrostatic model contributed to a better understanding of many experimental results. I shall not discuss here physical properties like melting and boiling temperatures, compressibility, thermal expansion, hardness and others. But I would like to discuss in some more detail the values of the enthalpies of formation.

1. It is well known that the enthalpy of formation of the iodide of a given element is lower than that of the fluoride. As the values for the electron affinity minus half the dissociation energy are not very different for the different halogens and as the other values of the cycle are the same, the difference of the enthalpy of formation is a simple consequence of the difference of the distances in the lattice.

2. Kossel has drawn attention to the fact that in salt-like compounds of the main groups the ions strive to attain the rare gas configuration. This is easy to understand for the halogen atoms, because the addition of one electron is an exothermic reaction. But the formation of positive ions and of negative ions with a double or triple charge takes place by endothermic reactions; the energy required must be delivered by the lattice energy.

In order to understand the rules of Kossel, Grimm and Herzfeld have calculated the enthalpy of formation of unknown compounds like NeCl or CaCl. Table 4 demonstrates that one gets positive enthalpies for such compounds or that a compound like CaCl is unstable compared with a mixture of $Ca + CaCl_2$.

Table 4

Enthalpy for formation of some halides according to calculations of Grimm and Herzfeld ($kcal/mole^{-1}$)

NeCl	+254	NaCl	- 98.5	MgCl	-18	$MgCl_2$	-151.0
ArCl	+126	KCl	-104.1	CaCl	-52	$CaCl_2$	-190.4
KrCl	+ 95	RbCl	-105.0	SrCl	-57	$SrCl_2$	-195.7

Of course this result is only valid if the unknown compounds
have salt-like structures. For NeCl with covalent bonds or
Ca_2Cl_2 with $[Ca-Ca]^{2+}$-ions other values are obtained.

3. The potential of ionization (ΔH_{IP}) is proportional to
   $1/r_A$ (r_A = radius of the atom, forming the cation A^+), whereas
   the lattice-energy (ΔH_U) is proportional to $\dfrac{1}{r_{A+} + r_{B-}}$. As
   the sum $r_{A+} + r_{B-}$ is always larger than r_A, it is seen
   that the most important factor for the stability of salt-like
   compounds is the potential of ionization; this is demonstrated
   by fig. 2.

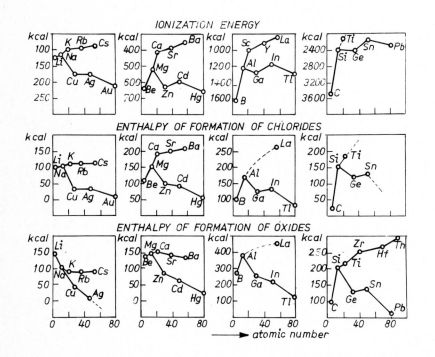

Fig. 2. Ionization Energy and Enthalpy of Formation, according
        to Klemm.

4. In some cases very high values of the energy of sublimation
   of the metals can be of importance. This is the reason why
   compounds of low valency of the transition metals are in
   general not stable.

5. Sometimes the calculations of the lattice energy must be
   modified. This will be discussed later on.

III. DIPOLES AND POLARIZATION

The attraction of permanent *electric dipoles* by ions and
the mutual interaction of dipoles belongs to any electrostatic
theory. Molecules with a dipole are attracted by ions; examples
will be discussed later on. Furthermore, molecules with permanent
dipoles attract one another more strongly than comparable molecules without dipoles. Therefore the boiling points are increased.
Van Arkel investigated this influence in detail; fig. 3 gives an
example of his results; no explanation is needed.

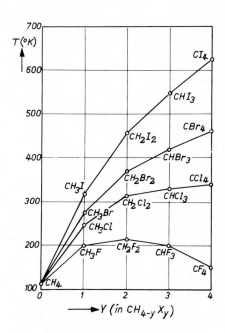

Fig. 3. Boiling points of some carbon
compounds, according to van Arkel.

Whereas it was clear from the very beginning that an electrostatic theory must consider the influence of permanent dipoles,
it is an inevitable consequence of such a theory that by
polarization of atoms and ions in an electric field dipoles
are induced, the strength of which depends on the polarizibility
of the species in question. In this way ions attract molecules
having a permanent dipole moment as well as molecules not having
such a moment.

In coordination lattices the influence of the polarization
is relatively small because the ions are surrounded by ions of

opposite charge in a symmetrical way, so that the effects com-
pensate each other; the remaining effects will be discussed later.
But in many single molecules and in molecule lattices such com-
pensation does not take place and therefore the polarization can
become of great importance. Fig. 4 demonstrates the situation
for a molecule like NaCl.

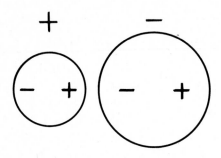

Fig. 4. Polarization in molecules like NaCl.

In this case the contribution of the polarization to the molecular
energy is not high, considering the low charge of the cation. The
effect is much more important for molecules with very small
cations which have a high charge, e.g. $SiCl_4$. Van Arkel and
de Boer and later on the author have made calculations for such
molecules. Fig. 5 gives the results.

One sees that the values of the molecular energy are correct
for the silicon halides and that even for the carbon halides
the agreement is not too bad. Here the energy of polarization
amounts to about 40% of the total molecular energy! It seems
important to me that neither in the experimental nor in the calculated values there is a jump indicating a change of the type
of bond.

The H^+-ion is a special case because it is a proton without
an electron shell. Therefore the proton will enter into the
electron shell of an anion until it is repelled by the positively charged nucleus of the anion. In this case the energy
as well as the shape of the molecule will be changed.
Let us consider H_2O, NH_3 and CH_4.

Fig. 6 demonstrates the forces in a water molecule.

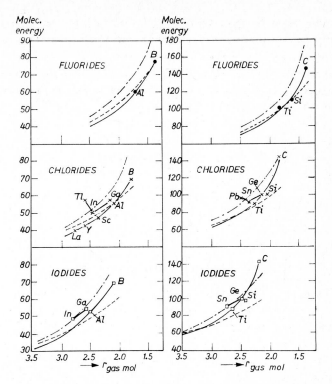

Fig. 5. Comparison of calculated and experimental molecular
energies, according to Klemm. $(kcal/mole^{-1})$
a) trihalides   b) tetrahalides
$\Delta H_{M,exp}$ gas mol. ——— $\Delta H_{M,theor}$ gas mol. —·—·—·—·
ΔH_u coordination lattice: ------- $\Delta H_u = \dfrac{a' \cdot e^2 \cdot N}{r_{cat}+r_{an}}$
r_{gas} molecule for trihalides $= 0.90 \left(r_{cat}+r_{an}\right)$
                   for tetrahalides $= 0.94 \left(r_{cat}+r_{an}\right)$
As the energy of repulsion is not considered, the calculated values are ca. 10% too high

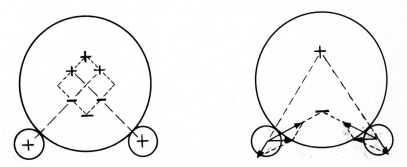

Fig. 6. Induced dipole moments in a water molecule.

The triangular shape is easy to understand considering the
polarization. Calculation gives the correct value of the molecular
energy. The same is true for ammonia.

However, one gets a wrong shape if one makes calculations
for methane on the basis of one C^{4-} and four H^+; the calculation
yields a tetragonal pyramid. In order to obtain the correct
tetrahedral shape, one must assume one C^{4+} and four H^-. The
situation with the NH^{4+}-ion is similar. In analogy to NH_3 one
should make use of the scheme: N^{3-} + four H^+; in this case the
NH_4^+-ion would have a pyramidal structure. The NH_4^+-ion is known
only in crystal lattices; there it has a tetrahedral structure.
With examples like CH_4 and NH_4^+ one has doubtlessly overextended
the applicability of the electrostatic model; the influence of
covalent bonds becomes decisive. The same may be true for PCl_3
or ClO_3^-; the pyramidal structure of such particles can be
explained by the fact, that the cation has two s-electrons
which produce a high polarizibility. But here too one must be
cautious; the influence of covalent bonds may be decisive in
determining the shape.

IV. SALT-LIKE COMPOUNDS

The field in which the electrostatic theory has its greatest
success are *crystals of salt-like compounds* the structure of
which could by no means be understood on the basis of the older
theories. Fundamental data necessary for the understanding of
crystal structures are the *radii* of atoms and ions. Values
for the radii of ions have been derived by
Goldschmidt, Pauling and others. The experience demonstrates
that the radii are roughly additive, this is extremely helpful
for practical reasons. But theoretically the shape of the curve
obtained for the radii can not be understood easily. If one
supposes similar attraction forces, the radii of ions with the
same electron configuration should decrease monotonically with
increasing nuclear charge. Fig. 7 reminds us that in reality
the shape of the curves is quite different. This is without
doubt due to the effect of the differences of the attraction
and the repulsion forces. Zachariasen and others have made
calculations on the basis of electrostatic models in order to
come to a better understanding of the shape of the curves. I
will not discuss here the results in detail, but only mention
that the important fact that the distances become shorter with

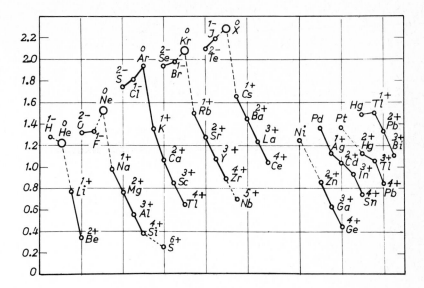

Fig. 7. Radii of the ions according to Goldschmidt

decreasing coordination number can be understood as the consequence of the decrease of the repulsion potential.

In a series of excellent investigations V.M. Goldschmidt
elaborated the general points of view in order to understand
the appearence of the different crystal structures. For AB_2-
compounds fig. 8 shows that there are two main influences: the
ratio of the radii and the *polarization*. The fact, that under
certain conditions layer-structures or even molecule-structures
are formed can be understood on the basis of an electrostatic
model. In coordination-lattices the influence of polarization
is low; in layer- and molecule-lattices it can become the deciding
factor.

In general the polarization effects are greater for anions,
but they can become important for cations too,if their polarizibility is high; this is important for the layer-structures
of Cs_2O, PbO etc.

As already mentioned polarization effects are in general
small in coordination-lattices; but here again they can have some
influence under two conditions:

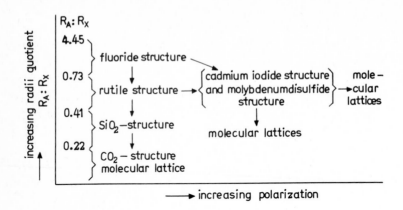

Fig. 8. The conditions for the appearance of the various types of crystal structures of AB_2-compounds, according to Goldschmidt.

1. If the polarizibility of the anion is high and if the cation has a rather high electronegativity, e.g. AgCl, the electron clouds of the anions are deformed in the direction of the cation; in other words, there is a transition to the covalent bond, i.e. the distances are decreased, the lattice energy is increased. In addition this explains why under these conditions structures having the coordination number 4 become predominant.

2. Another consequence of the polarization are the van der Waals-forces. The theoretical aspects of this type of bonding belong to quantum theory and cannot be discussed here in detail. In general the contribution of the van der Waals-forces to the lattice energy is small, but the influence of these forces can no longer be neglected if the polarizibility of both kinds of ions is high (e.g. AgI). These forces are the reason why CsCl, CsBr and CsI do not crystallize in the NaCl-, but rather in the more compact CsCl-structure and they are responsible for the occurrence of the $PbCl_2$-type in $BaCl_2$, $BaBr_2$ and BaI_2 and some halides of the rare earth elements.

V. STABILITY OF COMPLEXES

As mentioned earlier Kossel has demonstrated that the electrostatic theory provides an explanation for the formation

of *complexes* like BF_4^- and SO_4^{2-}. But one must bear in mind that
the calculations of Kossel were made for isolated gaseous
particles; in reality complexes of this kind are always combined
with cations in a lattice or in solution. In order to calculate
the enthalpy of formation of coordination compounds it is
necessary to make an energy balance considering the lattice- (or
molecular-) energies of the components, the energy of formation
of the complex and the lattice energy of the coordination
compound.

In this way the energy of the complex calculated by methods like
those of Kossel is only part of the total balance; even with high
values for the formation of the complex the total enthalpy of
formation can be small or even positive.

There are not enough systematic investigations available
in order to derive general rules for the stability of complexes
in the solid state. May I mention the work of W. Biltz on the
addition of ammonia on halide compounds. The alkali halides form a
few rather unstable ammonia compounds; in the second group the
ammonia compounds are much more numerous and more stable. The
stability decreases as the size of the cation increases, it
increases with increasing anion size (see Table 5). The second
effect is easy to understand as the lattice energy decreases
with increasing distance.

Table 5

Hexa-ammines of alkaline earth halides.

	Enthalpy of formation of the compound kcal	Energy for enlarging the lattice kcal	Change of energy on embedding six NH_3 in the enlarged lattice kcal
$CaCl_2$	-73	+100	-173
$CaBr_2$	-83	+ 96	-179
CaI_2	-95	+ 87	-182
$SrCl_2$	-60	+ 99	-159
SrB_2	-74	+ 95	-169
SrI_2	-84	+ 88	-172
$BaCl_2$	-54	+ 98	-152
$BaBr_2$	-63	+ 93	-156
BaI_2	-71	+ 87	-158

That the stability is inverted for the silver halides demonstrates
once more the strong polarization effects in these compounds,
these effects increase from the fluoride to the iodide. The
dependence on the size of the cation can be explained by the
fact that the energy of attraction of a dipole by an ion depends
on $1/r^2$, whereas the energy of the lattice depends on $1/r$.

The stability of ion-dipole complexes depends on the size
of the permanent dipole moment and also on the induced moment;
so it can be easily understood why e.g. the ammonia complex
of the Ag^+-ion is more stable than the water complex.

Rather early one became aware of the fact that in the group
of the transition elements the same central ion can form two
classes of complexes which have different magnetic moments *high
spin*' and '*low spin*' complexes. In principle, in all complexes the
electron cloud of the ligands is strongly deformed in the
direction of the central ion. According to Fajans such a
deformation means that a transition to a covalent bond takes
place. This happens very often, but with complexes of the

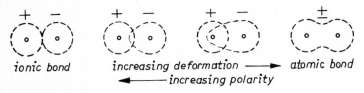

Fig. 9. Transition from an ionic to a covalent bond

transition elements the two forms - ion-dipole-electrostatic
bond and covalent bond - can be clearly distinguished. Later on
Pauling presented a detailed calculation of the magnetic moments
of complexes with covalent bonds. According to a very useful
rule octahedral and tetrahedral complexes are diamagnetic if the
total number of electrons surrounding the central ion equals
the number of electrons in a rare gas. (Square complexes with
the coordination number four are diamagnetic if the number of
electrons is smaller than that in a noble gas by two, e.g. in
the $[Ni(CN)_4]^{2-}$-complex.) The question of whether a complex has
ion-dipole-electrostatic or covalent bonds can be rather well
understood if one considers the strength of the electric field
of the central ion and the polarizibility of the ligands.

Later on Bethe and others demonstrated that a pure electrostatic model also enables us to understand the existence of
low and high spin complexes, since the splitting of d- or
f-electron levels in an electric field depends on the symmetry
and the strength of the field. It was specially impressing to
me that Hartmann demonstrated the diamagnetism of complexes like
$[Ni(CN)_4]^{2-}$ to be a consequence of the splitting of the d-
electron levels in the electric field. The modern development
has tended to unite both models.

VI. SOLUBILITY AND HYDRATION

In order to explain the fact that so many salts are soluble
in water, Kossel assumed that as a consequence of the high
dielectric constant of water the attraction between the ions
is decreased so much that the thermal energy is sufficient to
separate the ions. This idea is of course too primitive,
especially for concentrated solutions. The deciding factor is
the formation of complexes of the ions with water, i.e. the
hydration The hydration energy for the ions of a compound is
given by the sum of the lattice energy and the enthalpy of
solution. It is certainly a difficult problem to separate the
value of the total hydration energy into the values for the
single ions; but it is satisfactory that electrostatic calculations of the energy of formation of water complexes of ions
yield values which are in good agreement with measured values.
Furthermore it is possible to derive some rules for the
solubility of salt-like compounds, although one must admit that
these rules are still rather rough.

The question of whether hydroxides split off H^+- or OH^--ions
more easily in aqueous solution has been discussed by Kossel.
The results of these calculations can be found in any textbook.
Actually, these calculations are only valid for the gaseous
state; in solution additional factors, especially the formation
of OH_3^+ and the hydration of OH_3^+ and the anions, play an important
role. But van Arkel and de Boer showed that the results of
Kossel are not changed by these effects in principle - on the
contrary, the contrasts become larger.

VII. FINAL REMARKS

It is not possible to discuss many other applications of
the electrostatic theory in this contribution. In closing,
a survey of all types of chemical bonding is given in fig. 10.

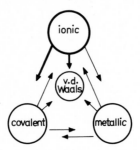

Fig. 10. Types of Chemical Bonding

The field especially treated in the books of van Arkel is drawn
in bold lines. Today one prefers in many cases to proceed from
other models, especially from the covalent bond - although the
calculations thereby become much more cumbersome. The fascinating
advantage of electrostatic models is the fact that it is rather
easy to make calculations the results of which are in many cases
in good agreement with measured values. But of course most of
these models must be modified to take into account the influence
of other types of bonding. Some papers written later by Prof. van
Arkel are excellent examples of how this should be done. I
believe that the beginnings of nearly all the problems which
will be discussed at this symposium can be found in the papers
of Prof.dr. A.E. van Arkel.

Crystal Structure and Chemical Bonding in Inorganic Chemistry
Eds. C.J.M. Rooymans and A. Rabenau
© 1975, North-Holland Publishing Company, The Netherlands

MAGNETIC PROPERTIES AND CHEMICAL BONDING

P.F. Bongers
Philips Research Laboratories
Eindhoven, The Netherlands

I. INTRODUCTION

In many compounds of transition-metal elements the metal ions
have a magnetic moment. These moments originate from the partly
filled 3d, 4d, 5d shell or in the case of the rare earth elements
the 4f electron shell. Although these electrons are the outer
electrons of the cations, they do not participate strongly in the
covalent bonding. The major part of the covalent bonding energy is
determined by the outer electrons of the anion which form bonds
with the cations by mixing with the empty s- and p-shells of the
cation. In the case of the first row-transition-metal sulphides,
the filled sulphur 3p shell mixes with the empty 4s and 4p
cation orbitals to form bonding states which, in a periodic
lattice, combine to the valence band. Therefore the information
one can obtain about the bonding character from the magnetism of
transition metal compounds is limited. However, a number of
magnetic phenomena give at least an indication of the bonding
character.

We will subsequently discuss the magnetic exchange interaction
in the class of compounds which are considered to be ionic. Then
we will describe a series of compounds in which the cations lose
their magnetic moment because of cation-anion covalency effects.
Finally we will present some cases where the cations have lost
their magnetic moment through cation-cation bonding.

II. EXCHANGE INTERACTION IN IONIC COMPOUNDS

In most of the fluorides, chlorides, oxides, etc. of metal
ions, which belong to the first row of transition metals, these
ions have magnetic moments which are spin-only moments. The
complete, or nearly complete, quenching of the orbital momentum
can be described readily by a picture of completely ionic bonding

if one takes into account the magnitude and symmetry of the
electrostatic interaction with the neighbouring ions which together
generate the crystal field. However, the magnetic interaction
between magnetic ions cannot be described by a purely ionic
picture. This is illustrated in Table 1 where some examples are
given of the magnetic ordering temperatures T_N of coupounds
containing cations with five 3d electrons. All these compounds
have a perowskite or perowskite-like structure and the spatial
arrangement of the magnetic cations and the anions is the same for
all compounds. The value of T_N increases as the difference in

Table 1

		$T_N(K)$		$T_N(K)$
a. d^5 ions	$Fe^{3+}F_3^-$ $RbMn^{2+}F_3$	394 82	$LaFe^{3+}O_3^{2-}$	750
		$\theta_p(K)$		$\theta_p(K)$
b. d^3 ions	$CrCl_3$ $CrBr_3$ CrI_3	+ 30 + 41 + 70	$NaCrO_2$ $NaCrS_2$ $NaCrSe_2$	- 354 + 30 + 130

The Neel temperature T_N and the asymptotic Curie temperature θ_p
of several halides and chalcogenides of d^5 and d^3 ions.

a. compounds with perowskite(-like) structures $AMeX_3$; the angle
   Me-X-Me is 180 degrees

b. CrX_3 and $NaCrX_2$ compounds; the angle Cr-X-Cr is 90 degrees

charge between cation and anion increases. The electronegativity
also plays a role in the magnetic interaction, as could be inferred from Table 1 . Here the asymptotic Curie temperatures
θ_p of the Cr^{3+} halides are listed. θ_p is determined from the
Curie-Weiss behaviour of the susceptibility $\chi = \dfrac{C}{T- \theta_p}$ and θ_p
is proportional to the sum of all magnetic interactions between
neighbouring cations. Note that θ_p increases in spite of the
increasing cation-cation distance in the sequence from chloride,
to bromide to iodide. Of course these conclusions based on a few
series of compounds are only qualitative and they are intended to
give an indication of the kind of parameters that are involved in
the exchange interaction.

The sign and magnitude of the magnetic interaction which is
found in the transition metal fluorides and oxides [1] can be
described very well, but qualitatively, however, by the theory of
Anderson [2] and the Goodenough/Kanamori exchange rules [3,4]. The
essence of the Anderson model can be explained with the aid of
fig. 1. If the d-orbitals of two neighbouring cations each containing one electron, overlap with a p-orbital of the anion in between,
two effects are of importance.

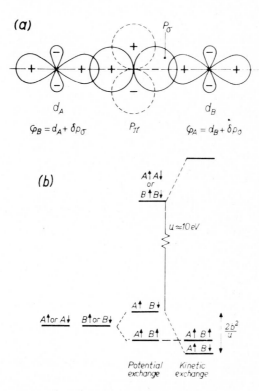

Fig. 1 a. Overlap of d_{z^2} orbitals of two cations with the p-
        orbitals of one anion.
      b. Schematic energy level diagram of parallel and anti-
        parallel orientation of the spins of two cations
        separated by an anion.

As a consequence of this overlap the orbitals of the cations
have some anion p-character and the unpaired electrons meet at
the anion. This leads to a lowering of the energy of the
configuration with parallel spins. This is essentially the same

mechanism that gives a parallel alignment of the spins of the
electrons at one atom according to Hund's rule. The second effect
involves the transfer of the electron of a cation to the other one
which means mixing-in a state of very much higher energy (U higher)
which has both electrons at ion A or both at cation B in the same
orbital (see fig. 1). Because of the Pauli principle this is only
possible with antiparallel spins. Only the state with antiparallel
spins is lowered (by b^2/U, where b is a measure of the overlap).
Usually this latter effect is larger which means that the total
interaction is antiferromagnetic. However, if the overlap is
zero by symmetry, so that no transfer is possible, as is the case
for the cation-anion-cation configuration over a 90° angle, as
given schematically in fig. 2a, the ferromagnetic interaction
dominates.

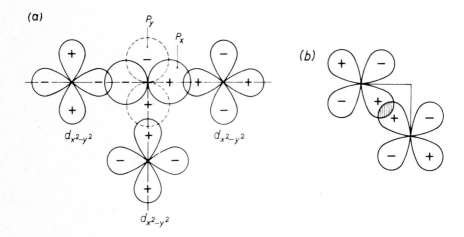

Fig. 2 a. Overlap of the $d_{x^2-y^2}$ orbitals of three cations with
          the p-orbitals of an anion. Overlap with the same p-
          orbital leads to a negative contribution to the inter-
          action between cations. Overlap with different p-
          orbitals gives a positive contribution to the interaction
          between cations.
       b. Direct overlap of d_{xy} orbitals of cations at a short
          distance.

By considering the symmetry of the crystal field, the number
of filled, half-filled and empty d-orbitals and the possible
transfer via overlap, Goodenough and Kanamori [3,4] have set up

a series of rules for the magnetic exchange interaction between
metal ions in various configurations.

Direct overlap of d-orbitals is also possible, as is
demonstrated in fig. 2b. This direct overlap, which is very
dependent on cation-cation distance, is the reason why, for
instance, $NaCrO_2$ is antiferromagnetic whereas the average interaction between the Cr^{3+} ions in the corresponding sulphides and
selenides and also in the CrX_3 halides - where the Cr-Cr
distances are larger - is ferromagnetic (see table 1).

Using the Goodenough/Kanamori rules and the concept of
direct overlap it has been common practice in the past years to
"explain" succesfully the sign and the magnitude of the magnetic
interactions of the majority of the insulating transition metal
halides and oxides and sulphides. Especially the interaction
between Cr^{3+} ions has been studied. One of the most extensive
analyses of the $Cr^{3+}-Cr^{3+}$ interactions as determined from susceptibility and other magnetic data has been given by Motida
and Miyahara [5] . In fig. 3 the asymptotic Curie temperatures of

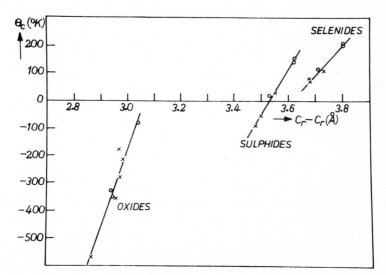

Fig. 3 Asymptotic Curie temperature θ_p as a function of the Cr^{3+}-
Cr^{3+} distance for compounds in which the nearest neighbour
Cr^{3+}-anion-Cr^{3+} angle is 90 degrees (from ref. 7 based on
data given in ref. 5).

various compounds are shown as a function of the $Cr^{3+}-Cr^{3+}$ distance. The increase of the asymptotic Curie temperature with

$Cr^{3+}-Cr^{3+}$ distance is attributed to the positive contribution of
the exchange via the anion and a negative interaction caused by
the direct overlap between the d-electron orbitals of neighbouring
chromium ions (see fig. 2b). The latter interaction dominates at
short cation distances. Thus, according to the models outlined
above, the magnetic properties indicate that at short cation-
cation distances a direct interaction between the d-electrons is
of importance.

Recently a series of elegant physical experiments has shown
that for the interaction between Cr^{3+} ions the model is correct.
Henning et al. [6] found from the E.S.R. of pairs of Cr^{3+} ions
in $ZnGa_2O_4$ spinel that the interaction between nearest neighbours
is very much larger than between more distant neighbours. Van
Gorkom et al. [7] studied the optical absorption, fluorescence
and excitation spectra of these pairs of ions. Their results
confirm the qualitative picture sketched above, giving positive
interactions between spins of electrons of non-overlapping
orbitals and negative interactions if the orbitals do overlap.

In the fluorides and oxides the transfer energy U (fig. 1)
is quite large. For NiO the value of U is quoted as approximately
7 eV [1] and recent measurements on $Y_3Fe_5^{3+}O_{12}$ indicate that U is
3 eV in this compound [13a]. In the following sections we will
concentrate on two series of compounds in which the transferred
state has a very low energy.

III. THE PYRITES

The disulphides of the 3d transition metal ions have the
pyrite structure. The cations are divalent and the sulphur atoms
form S_2^{2-} groups. The structure is presented in fig. 4. The
cations are octahedrally surrounded by six sulphur atoms. The
magnetic and electric properties of the compounds MeS_2 with
Me = Mn, Fe, Co, Ni, Cu are reviewed in table 2. A great variety
of properties has been observed. Some of the compounds are
metallic conductors, others insulators while furthermore paramagnetic; ferromagnetic as well as nonmagnetic behaviour has been
observed. The pyrites and the series of mixed crystals between
these compounds have been studied extensively by Bither, Bouchard,
Jarret and others [8,9,10]. We shall explain the properties of
the isolated pyrites with the aid of the data and models presented

Table 2

	MnS_2	FeS_2	CoS_2	NiS_2	CuS_2
n_d	5	6	7	8	9
magn. prop.	AF	TIP	F	AF	PP
n	5	0	0.8	2	0
el. cond.	s.c.	s.c.	m.	s.c.	m.

The properties of transition metal pyrites MeS_2:n_d is the number of
d-electrons per Me^{2+}; antiferromagnetic AF, Pauli paramagnetic
PP, temperature-independent paramagnetic TIP, ferromagnetic F.
n denotes the number of unpaired electrons as obtained from
susceptibility or magnetization data.
semiconductor: s.c. and metallic conduction: m.

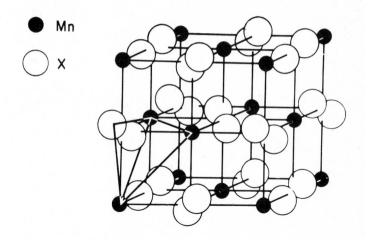

Mn

X

Fig. 4 The pyrite structure, MeX_2.

by these authors, using a molecular orbital diagram given for
FeS_2 similar to that presented by Goodenough [11] (see fig. 5).
    MnS_2 (Mn^{2+}:d^5) which is an insulator and is paramagnetic
with S=5/2 at room temperature and antiferromagnetic below
T_N = 78K; it behaves as an ionic compound. The five 3d electrons
are in the three t_{2g} and the two e_g orbitals. The t_{2g} orbitals
have little overlap with the anion orbitals and are considered to
be localized and non-bonding. The e_g orbitals are directed towards
the anions and the overlap with the 3p orbitals shifts the anti-

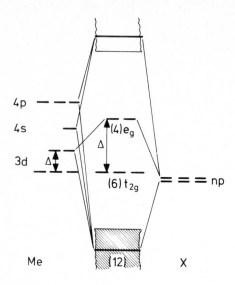

<u>Fig. 5</u> Molecular orbital (MO) energy level diagram of a 3d cation
         surrounded by six sulphur atoms. The bonding levels together
         form the valence band (schematically indicated). The anti-
         bonding levels formed by $4s$ and $4p$ of the cation and $4p_s$ of
         the sulphur atoms constitute the valence band. The anti-
         bonding levels of the e_g and sulphur $4p$ orbitals form a
         narrow d-band or localized levels as discussed in the text.

bonding e_g orbitals to higher energy. Also the spin pairing energy
is decreased because the unpaired electrons are spread more in the
direction of the anions.

The Fe^{2+} ion has the d^6 configuration and one might expect
four unpaired electrons. This is not the case with the Fe^{2+} ions
in FeS_2 which is a semiconductor and the iron ions have no magnetic
moment. Apparently the antibonding e_g levels are shifted upwards,
far enough to cause the low spin state with all six electrons in
the t_{2g} levels to have the lower energy.

This is not unexpected as it is well known that Co^{3+}, (also
d^6) is in the low spin state in most of the oxides.

If we now turn to CoS_2 (Co^{2+}; d^7) it is found that this
compound is ferromagnetic with T_c = 115 K and the saturation
magnetization corresponds to 0.8 unpaired electrons per cobalt.
Therefore it is probable that the extra electron is in the e_g
level, and Co^{2+} is t_{2g}^6, e_g^1 in the low spin state, as is found for
Ni^{3+}; d^7 in $NaNiO_2$ [19]. The metallic conductivity of CoS_2
seems to indicate that the e_g electrons are non-localized,

meaning that the transfer energy U, discussed in the previous
section, is very small.

   Molecular orbital diagrams like fig. 5, are not suitable
for indicating the energy required for the transfer of an electron
from one cation to another. This can better be discussed with the
aid of a plot of the energy versus the density of states, as given
in fig. 6 for a compound with cations containing n d-electrons.
A convenient example is V^{3+} which has two d-electrons. The highest
filled states in fig. 6a denote the energy of the d-electrons of
the V^{3+} ions. If an electron is transferred from one ion to
another then the third electron gets an energy d^{n+1}. If now the

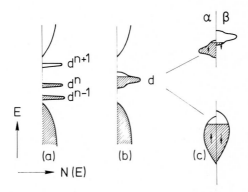

<u>Fig.</u> 6 Energy versus density of states for a compound containing
        cation; hatched area denotes filled electron states with
        n d-electrons (d^n)
        a. energy levels of electrons of cations with d^n, d^{n-1} and
           d^{n+1} configurations are given. The electrons are
           localized.
        b. the d-electrons are in a narrow partly filled d-band.
        c. spin polarization of a narrow band (top) and a broader
           band (bottom).

the transfer energy overlap increases, the levels become broader,
and the transfer energy (U) is decreased. The peaks of the density
of states become less separated and finally merge into a narrow
d-band which is partially filled (fig. 6b). If this band is narrow,
it can be appreciably spin-polarized by the exchange energy but
much less if the band is broad. This is illustrated for the two
cases in fig. 6c.

   If we return to the pyrites it is apparent that the combination of metallic properties and ferromagnetism of CoS_2 with nearly one unpaired electron per cobalt fits the description of a well polarized narrow e_g band which is one quarter filled. CuS_2 is diamagnetic and metallic: here the covalency is expected to be stronger so that the e_g band, which is three quarter filled, is broader. This might explain why CuS_2 is not magnetic. On the other hand, there is evidence that in all Cu^{2+} sulphides the copper d-band is below the top of the valence band, causing copper to be monovalent and the metallic conduction to occur through holes in the valence band.

NiS_2 forms a special case. It is semiconducting and the nickel ions have two unpaired (e_g) electrons. In this compound the e_g band is half-filled and the electrons are apparently localized. Localization is related to electron-electron repulsion and may be particularly strong in case of a half-filled narrow band.

As has been described by Mott [12] , such an ordered state can transform to a metallic state by a first-order transition. Mott's arguments are: If the electrons are ordered by electron-

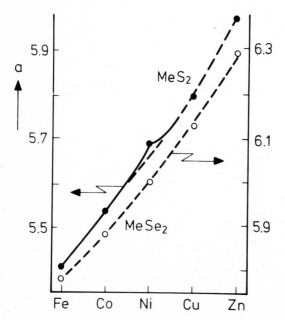

Fig. 7 Lattice parameter of transition metal pyrites. Sulphides: left-hand scale; selenides: right-hand scale. The drawn line refers to measurements on mixed compounds (see ref. 13).

-electron repulsion it requires the correlation energy for one
electron to become itinerant. The more electrons become
itinerant, the smaller the excitation energy becomes, because
the itinerant electrons have a screening effect on the localized
electrons. Thus leading to a cooperative transition.

In fig. 7 it is shown that the lattice constant of the semi-
conducting NiS_2 is out of line with the other disulphides. This
is not the case for the corresponding selenides which are all
metallic. Bouchard et al. [10] have found that in the series
$NiS_{2-x}Se_x$ compositions occur which indeed show an insulator-to-
-metal transition (see fig. 8).

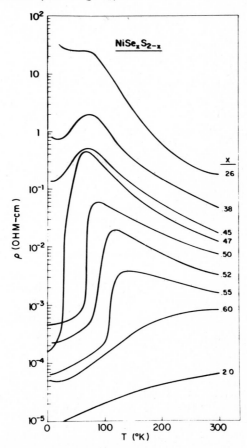

<u>Fig. 8</u> Resistivity versus temperature of crystals of $NiS_{2-x}Se_x$
(according to ref. 10).

In the foregoing we have shown that the covalent bonding between the cation outer electrons and the anions can lead to metallic conduction and the loss of magnetic moments. The properties can be described with a simple concept of the band structure of these compounds. It should be noted that in the present description of the pyrites the valence bond is assumed to have the normal quadratic band form. Possible deviations of this band form caused by the formation of S_2^{2-} are not taken into account and might well be of importance. Direct measurements of the density of states by methods such as ESCA, Auger spectroscopy, etc. are clearly needed to check the qualitative description given above.

IV. EFFECT OF CATION-CATION BONDING

In this section we shall discuss some examples where d-electron bands are formed by direct overlap between the d-orbitals. In contrast to the cation-anion covalency which increases if one moves to the right in the periodic table (because of the incomplete screening by the d-electrons of the cation nucleus the anion electrons are pulled more to the cation), the direct d-d overlap increases if one moves to the left of the periodic table. This is because the d-orbitals are less contracted at the beginning of the transition metal series. Also ions with a small number of d-electrons, octahedrally surrounded by anions, have their electrons in t_{2g} orbitals which extend mainly between the anions. Therefore it is best to look for structures where the cations are in octahedra which share edges. This is the case for the octahedral coordinated sites in the spinel structure.

In table 3 some of the properties of $CuCr_2X_4$ spinels are listed. $CuCr_2O_4$ is an insulating ferrimagnet of divalent copper and trivalent chromium. $CuCr_2S_4$ is metallic and ferromagnetic

Table 3

$CuCr_2O_4$	ferrimagnetic	$T_N = 90$ K	insulator
$CuCr_2S_4$	ferromagnetic	$T_C = 420$ K	metal
CuV_2S_4	Pauli paramagnetic		metal
$CuTi_2S_4$	Pauli paramagnetic		metal

with no magnetic moment at the copper site. The corresponding
vanadium and titanium sulphospinels are metallic and not one of
the metal ions has a magnetic moment.

Lotgering and van Stapele [14] have explained the properties
of $CuCr_2S_4$ in the following way. The copper d-levels are shifted
to lower energy with respect to the (sulphur) valence band (see
fig. 9), so that even the Cu^{1+} level is below the top of the
valence band. Copper becomes monovalent and is nonmagnetic and the
average valency of sulphur is less than S^{2-}. The chromium moments
are localized and the ferromagnetic interaction is enhanced by the

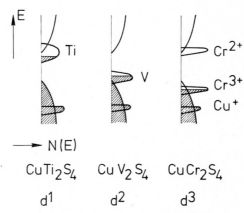

Fig. 9 Energy versus density of states of $CuTi_2S_4$, CuV_2S_4 and
$CuCr_2S_4$.

itinerant holes in the valence band (see fig. 9). Recently
Hollander et al. [15] confirmed the presence of Cu^{1+} in $CuCr_2S_4$
by measuring and comparing the ESCA spectra of $CuCr_2S_4$ and of
well known Cu^+ and Cu^{2+} compounds. In CuV_2S_4 the vanadium ions are
not magnetic and apparently the d-electrons are itinerant in a
narrow d-band. The existence of a narrow band is in agreement with
the high Pauli paramagnetic susceptibility found for this compound.
CuV_2S_4 is a superconductor at low temperatures. For copper titanium
sulphide the d-band is expected to be broader than in the vanadium
spinel.

The formation of d-bands in compounds with ions, with one or
two d-electrons in octahedral coordination, also occurs in several
oxides, especially in the vanadium oxides. The properties of the
lower oxides of vanadium have been studied intensively during the

last decade.

There is a large number of vanadium oxides with well-defined
composition between V_2O_3 and VO_2. These compounds form a series
of so-called Magnéli phases or shear structures whose chemical
formula is given by V_nO_{2n-1}. At high temperature, VO_2 has the
rutile structure. The V^{4+} ions are in octahedra which share
edges with some of the neighbours and corners with the other
neighbouring VO_6 groups. In V_2O_3 edge- and face-sharing occurs.
In the phases in between, corner-sharing has been subsequently
replaced by edge-sharing and some of the common edges have been
replaced by face-sharing.

Recently Kachi et al. [16] have investigated many of these
vanadium oxides. It turns out, as can be seen in fig. 10, that

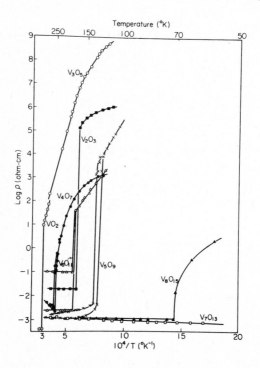

Fig. 10 Electrical resistivity versus reciprocal temperature of the
vanadium oxides V_nO_{2n-1} according to Kachi et al. ref. 16.

most of the compositions show insulator-to-metal transitions. In
the metallic phase all compounds show a paramagnetic Curie-Weiss
behaviour of the susceptibility with Curie constants in fair

agreement with the calculated value for the V^{4+} and V^{3+} combination under consideration. This leads to the conclusion that if the metallic conduction is caused by the d-electrons this band must be a narrow band.

Several of these compositions exhibit a maximum in the susceptibility temperature curve below the electrical transition temperature and it was verified by Mössbauer effect that these maxima coincide with the magnetic ordering temperature. Apparently the magnetic order has nothing to do with the insulator-metal transition. Whether the electrical transition is caused by a Mott transition, cation-cation bonding or an electron-phonon instability is in my opinion still a point at issue. In fig. 11 a survey is given of the electrical and magnetic transition temperatures. Only for V_2O_3 do the magnetic ordering temperature

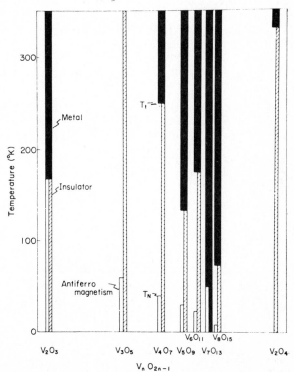

Fig. 11 Magnetic and electrical phase diagram of the compounds V_nO_{2n-1}; black: metal; hatched: insulator and dotted: antiferromagnetic (from ref. 16).

and the electrical transition coincide. The most interesting for
the present discussion is VO_2, since it seems that the most
likely explanation for the electric transition is the formation
of cation-cation bonds. In the insulating state, the V^{4+} ions in
face-sharing octahedra form pairs with a short V-V distance. No
magnetic moment could be detected for the V atoms within the pairs.

Another compound in which vanadium atoms seem to become nonmagnetic through cation-cation bonding is lithium vanadium oxide.
$LiVO_2$ has the $\alpha NaFeO_2$ structure which can be considered to be
an ordered rocksalt lattice with the (111) metal layers alter-
nately occupied by Li and V. The vanadium ions form a triangular
arrangement in the layers. At 450 K a drastic change in the
lattice parameter occurs [17]. At the transition the a-axis
expands and the c-axis contracts with increasing temperature.
A strong increase of the susceptibility occurs at the transition
as is shown in fig. 12. Above the transition the compound is

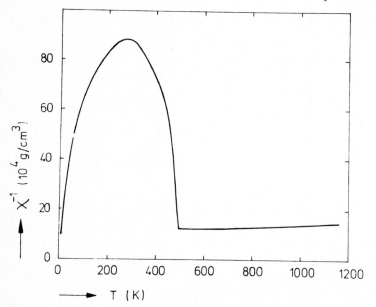

Fig. 12 Temperature dependence of the inverse magnetic susceptibility of $LiVO_2$. The decrease of χ^{-1} below 220 K is
attributed to 0.006 V^{4+} ions per formula unit.

paramagnetic and insulating and below the transition $LiVO_2$ is an
insulator too. Below the transition temperature the susceptibility

can be described by $\chi = \chi_o + C/T$, i.e. a small temperature-
-independent contribution and a paramagnetic contribution. The
latter is attributed to 0.004 V^{4+} per formula unit which agrees
well with the 0.006 V^{4+} determined in the sample by chemical
analysis.

Measurements by Locher [18] of the ^{51}V nuclear magnetic
resonance show a very low shift of 0.37%, which is essentially
temperature-independent between 2 K and the transition
temperature. This result clearly shows that the V^{3+} ions have no
magnetic electron spin moment. It is interesting that the NMR
results reveal that the local symmetry at the vanadium ion is
orthorhombic below T_t. In the $\alpha NaFeO_2$ structure the local
symmetry is trigonal. This result might be the first indication
that the model proposed by Goodenough [3] is correct. In this
model it is assumed that triangular clusters of cation-cation
bonded vanadium ions are formed, as shown in fig. 13.

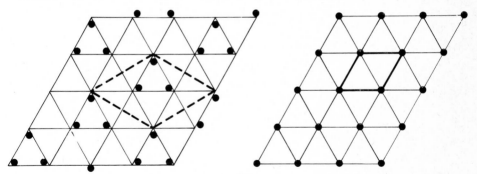

Fig. 13 Possible arrangement of V^{3+} ions in the (001) layers in
LiVO$_2$ (left) below 450 K and the triangular configuration
of the ions above 450 K ($\alpha NaFeO_2$ structure) at the right.

From X-ray powder data we have not been able to extablish a super-
-cell or a lower symmetry. Pure single crystals should be made
to verify this assumption.

V. CONCLUSION

In conclusion we can state that there are a number of
transition metal compounds in which the metal ions have lost their
magnetic moment by virtue of the covalent bonding involving the
anion and the d-electron orbitals. These effects occur mainly at

the right hand side of the transition metal series. Narrow d-band
formation and cation-cation bonding are the origin of the dis-
appearance of magnetic moments in compounds of ions with a small
number of d-electrons.

The models which were used to describe the properties of
series of compounds are crude and may be wrong in some respects.
More experiments to test the models, especially those which
directly determine the energy band scheme by spectroscopic
measurements, such as ESCA, photo emission, etc. are needed.

References

1   J. Owen and J.H.M. Thornly, Rep. Progr. Phys. 29 (1966) 675.
2   P.W. Anderson, Phys. Rev. 115 (1959) 2
    P.W. Anderson in: G.T. Rado and H. Suhl (eds.) Magnetism
    Vol. I, Academic Press, New York (1963),
3   J.B. Goodenough, Phys. Rev. 100 (1955) 564.
    see also J.B. Goodenough in: Magnetism and the Chemical Bond,
    Interscience, New York 1963.
4   J. Kanamori, J. Phys. Chem. Solids 10 (1959) 87.
5   K. Motida and S. Miyahara, J. Phys. Soc. Japan 29 (1970) 516.
6   J.C.M. Henning, J.H. den Boef and G.G.P. van Gorkom, Phys.
    Rev. B7 (1973) 1825.
7   G.G.P. van Gorkom, Thesis Utrecht, June 1973.
    G.G.P. van Gorkom, J.C.M. Henning and R.P. van Stapele, Phys.
    Rev. B8 (1973) 955.
8   T.A. Bither, R.J. Bouchard, W.H. Cloud, P.C. Donchere and
    W.J. Siemens, Inorg. Chem. 7 (1968) 2208.
9   H.S. Jarrett, W.H. Cloud, R.J. Bouchard, S.R. Butler, C.G.
    Frederick and J.L. Gillson, Phys. Rev. Lett. 21 (1970) 617.
10 R.J. Bouchard, J.L. Gillson and H.S. Jarrett, Mat. Res. Bull.
    8 (1973) 489.
11 J.B. Goodenough, J. Solid State Chem. 3 (1974) 26.
12 N.F. Mott, Proc. Phys. Soc. London A62 (1949) 416.
13 J.A. Wilson and G.D. Pitt, Phil. Mag. 23 (1971) 1297.
13a P.K. Larsen and R. Metselaar, J. Solid State Chem. 12 (1975)
    253.
14 F.K. Lotgering and R.P. van Stapele, J. Appl. Phys. 39
    (1968) 417.
15 J.C.Th. Hollander, G. Sawatzky, C. Haas, Solid State Comm.
    15 (1974) 747.
16 S. Kachi, K. Kosuye and H. Okinaka, J. Solid State Chem. 6
    (1973) 258.
17 P.F. Bongers, Chemisch Weekblad 63 (1967) 353
    and P.F. Bongers quoted by J.B. Goodenough, Magnetism and the
    Chemical Bond, Interscience New York, 1963, p. 270.
18 P.R. Locher and P.F. Bongers, to be published.
19 P.F. Bongers and U. Enz, Solid State Comm. 4 (1966) 153.

Crystal Structure and Chemical Bonding in Inorganic Chemistry
Eds. C.J.M. Rooymans and A. Rabenau
© 1975, North-Holland Publishing Company, The Netherlands

ALKALI METAL SUBOXIDES - A KIND OF ANTI-CLUSTER COMPOUNDS

by

Arndt Simon

Max Planck Institut für Festkörperforschung,

7 Stuttgart, Germany

SUMMARY

The alkali metals rubidium and cesium form suboxides MO_x with $0 < x < 0.5$. The compounds Rb_6O, Rb_9O_2 as well as Cs_7O, Cs_4O and $Cs_{11}O_3$ have been investigated by thermal analysis and x-ray methods and characterized by their crystal structures. These structures are discussed in terms of quasi-molecular groups of compositions Rb_9O_2 and $Cs_{11}O_3$, which are the only constituents of this class of compounds.

In the more metal-rich compounds these groups alternate with purely metallic structural parts as is indicated by the formulas $Cs_7O \,\hat{=}\, [Cs_{11}O_3]\,Cs_{10}$, $Cs_4O \,\hat{=}\, [Cs_{11}O_3]\,Cs$ and $Rb_6O \,\hat{=}\, [Rb_9O_2]Rb_3$.

I. INTRODUCTION

The alkali metals are located far away from the center of the periodic table but also far from the center of recent interest of chemists. Doubtlessly that is due to the strong tendency of these elements to occur as monovalent, rare-gas like ions. So chemistry of these metals seems to be colourless and simple. But one only needs to have a look at the oxides of the alkali metals to be aware of some unexpected complexity.

Table I summarizes all of the known alkali metal oxides with the formulas assigned to them at the beginning of our work in this field.

Besides the normal oxides of composition M_2O there exists a great number of oxygen-rich compounds [1]. These are due to the flexibility of oxygen, that may occur as O^{2-}, O_2^{2-}, O_2^- or even as an O_3^--ion. Moreover, some metal-rich compounds, the so-called suboxides, exist [2-7], and these are dealt with in our investigations [8-11].

Table I

O/M \ M:	Li	Na	K	Rb	Cs
3	–	NaO_3	KO_3	RbO_3	CsO_3
2	?	NaO_2	KO_2	RbO_2	CsO_2
1.5	–	–	–	Rb_2O_3	Cs_2O_3
1	Li_2O_2	Na_2O_2	K_2O_2	Rb_2O_2	Cs_2O_2
0.5	Li_2O	Na_2O	K_2O	Rb_2O	Cs_2O
< 0.5	–	–	–	–	Cs_3O
					Cs_7O_2
					Cs_4O
					Cs_7O

The chemistry of metals in low formal oxidation states has
been of considerable interest in recent time. It is mainly the
occurrence of the metal-metal bonding found with low-valent
compounds of transition metals (especially 4d- and 5d-elements
like Nb, Ta, Mo, W and Re) as well as with main group elements
like Bi or rare earth metals like Gd, that causes this interest.
To establish metal-metal bonds, compounds with a relatively high
metal content are needed, which means compounds with an excess
of electrons at the metal atoms capable of forming homo-nuclear
bonds due to lack in anion content.

The alkali metal suboxides have extremely high metal contents
as indicated by a formula like Cs_7O. Besides this, the structures
of sub-oxides turned out to be of a really surprising similarity
from a geometrical point of view to some of the structures of
the above mentioned transition metal compounds but in an inverse
way, which fact is outlined by the title of this contribution.

In the case of the transition metal compounds metal-metal
bonding may lead to the formation of discrete clusters as found
for example in compounds like $TaCl_{2.5}$ [12], molybdenum dihalides
[13], or the ternary $K_3W_2Cl_9$ [14]. Octahedra, or - in the latter
compound - pairs of metal atoms are the centers of the quasi-
molecular groups of atoms within such clusters. In other cases
metal-metal bonding may result in the formation of infinite
bonding regions as with $MoBr_3$ [15], the structure of which is
shown as a projection along [001] of the hexagonal cell in fig. 1.

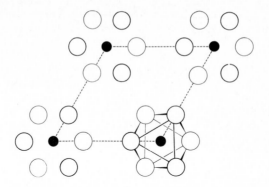

Fig. 1. $MoBr_3$, projection along [001]. (open circles: Br^-).
        Assumed structure for Cs_3O (open circles: Cs).

There is a hexagonal close packed arrangement of bromine atoms,
with the metal atoms occupying octahedral interstices. The
arrangement results in infinite chains of face-sharing coordination
octahedra parallel [001].

    If now cations and anions in $MoBr_3$ are interchanged, one
arrives at the anti-type of structure as assumed for the suboxide
Cs_3O [16]. The weak point in this assumption is the fact that
according to our investigations a stoichiometric compound of this
composition does not exist. The crystal structure is based on
data of a crystal with the presumed composition Cs_3O. Nevertheless
the inverse relationship between the structure of the low-valent
transition metal compound and the structure, though hypothetical,
of one of these alkali metal compounds has also been observed
for other suboxides. All alkali metal suboxides with the exception
of the still insufficiently described Cs_3O contain discrete atomic
clusters, but in these suboxides the centers of the clusters
are oxygen instead of metal atoms.

II. EXPERIMENTAL METHODS

    The cesium suboxides have been found early this century
by Rengade [2,3] who used thermal analysis as a method of investigation. The compounds were described in the textbooks with the
formulas of Table I, but despite of their unusual nature time has
passed with only a limited number of studies on them. Recently
the exact compositions, correcting some of the given formulas,

and an insight into the chemical bonding of this group of compounds has been established. Handling difficulties due to
extreme air-sensitivity as well as complicated and hardly
reproducible phase-relationships have delayed a deeper
knowledge of the alkali metal suboxides for a long time. The
difficulties could only be overcome by developing new techniques
for the handling and the study of these materials.

At first therefore an experimental outline of our own work
in this field is described. Table II gives the main path of our
investigations along with the methods used to elucidate special
aspects.

Table II

Investigations on the alkali metal suboxides

---

1. synthesis

$(M + O_2)$

2. thermal analysis   3. x-ray polycryst. samples
   special DTA (-50° to +200°C;   continuous Guinier camera
   resolution $\pm$ 0.01°C)   (-190° to +350°C; $\pm$0.02°C)

4. single crystal preparation

5. structure investigation

First comes the synthesis of samples by combining the
elements. No really severe problems are present at this stage,
if an exact amount of oxygen is reacted carefully with the metal
till at the end the gas is entirely absorbed $(p < 10^{-5}$ torr) [11].

Subsequently a thermal analysis is performed in a specially
designed apparatus. Some important conditions have to be met for
a reliable interpretation of the results of these thermal
analyses:

a) the temperatures need to be maintained and measured accurately
   within some hundredth of a degree and
b) the samples need to be annealed for long periods of time and
   - after equilibrium has been reached - investigated with increasing temperature only.

Besides thermal analyses phase investigations are done by
x-ray powder experiments. The extreme air-sensitivity of the
substances on the one side and the complexity of the diffraction

patterns on the other side called for a special x-ray technique
and have led to a combined Debeye-Scherrer and Guinier type of
camera [17]. This has resulted in a special camera design with
movable film [18], shown in fig. 2. The main points of this

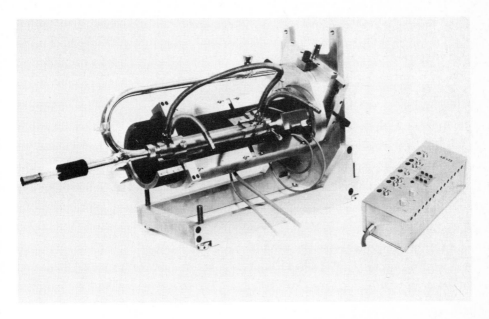

Fig. 2. Modified Guinier camera for temperature dependent studies
        [18].

instrument can be shortly summarized: the sample temperature can
be maintained accurately within some hundredth of a degree; the
heating and cooling rates can be made very small, smaller even
than half a degree per day. Thus there is the possibility of
investigating very narrow gaps between thermal effects such as
occur in the rubidium/oxygen system.

By these studies we get detailed informations on phase
relationships, compositions, reaction temperatures and - what
has turned out to be of extreme importance - kinetics of
formation for various compounds. We have thus been able to prepare
single crystals suitable for x-ray structure studies, the rest
being - as very often - more or less time-consuming routine
procedures.

III. CHARACTERIZATION OF ALKALI METAL SUBOXIDES

The results [11] concerning the compositions and the phase
relationships in the rubidium/oxygen and in the cesium/oxygen
systems are summarized in the figs. 3 and 4.

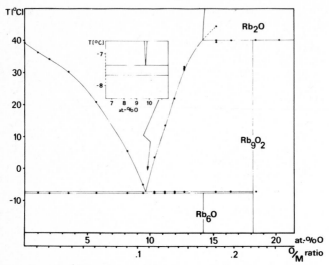

<u>Fig. 3</u>. Rb_2O/Rb phase diagram.

Because of the close neighbourhood of thermal effects in the
rubidium/oxygen system, the compositions of the existing phases
can only be determined roughly by thermal analyses and x-ray
powder studies. The exact stoichiometries have been derived by
single crystal investigations. Nevertheless thermal analyses
and x-ray powder data are extremely helpful to find the proper
conditions for getting those single crystals. These indicate
that the seemingly simple eutectic system is complicated by
the existence of an additional phase, Rb_6O, decomposing
practically at the melting temperature of the Rb/Rb_9O_2-eutectic.
Moreover, modified-Guinier photographs taken at heating rates
of two degrees per week right through the double effect gave clear
evidence of the order of the effects as drawn in the detail-
picture of fig. 3: Solid Rb_6O decomposes first with increasing
temperature leaving solid rubidium and solid Rb_9O_2 in the
intermediate phase field and three tenth of a degree higher
eutectic melting occurs. Subsequently rubidium or Rb_9O_2 melts
depending on the composition.

In the cesium oxygen system our own results substantiate
the beautiful work of Rengade [2,3] concerning the number of

existing compounds and their rough stoichiometries. But the
exact formulas partly deviate from those given by Rengade
especially in the oxygen-rich part of the diagram. So instead
of a stoichiometric Cs_3O there exists a broad homogeneity range,
the center of which is located at a higher oxygen content.
This makes a reinvestigation necessary of the structure determined
by Tsai and coworkers [16] for a crystal with, the only presumed,
composition Cs_3O (see p. 49).

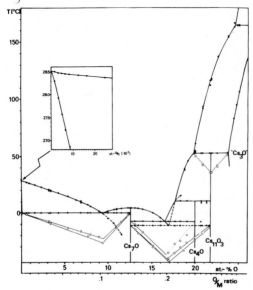

<u>Fig. 4</u>. The Cs_2O/Cs phase diagram
(• = temperatures, o = integrated DTA-peaks).

The permanganate-coloured compound, formulated as Cs_7O_2 by
Rengade, has to be described as $Cs_{11}O_3$. Composition and structure
of that special phase are of great importance in the description
of the alkali metal suboxides (see p. 59).

The investigation of the neighbouring phase Cs_4O has imposed
very tough problems. The difficulties originate in the fact that
this phase does not form and crystallize easily. Normally the
metastable eutectic with the neighbouring $Cs_{11}O_3$ forms. It needs
real tricks to teach the system appropriate behaviour. From our
experiments it is quite obvious that these difficulties are
caused by a low probability of the formation of seed crystals
of the compound Cs_4O. Consequently it happens to be impossible
to generate this phase within the tiny amounts of substance in
an x-ray capillary, a fact to be dealt with in a later section.

Cs_7O is the only compound, that melts congruently; it

crystallizes in bronze-coloured needles of hexagonal shape.
Again crystallization is delayed; so the liquidus reaches
deeply into the metastable region below the eutectic temperature,
but here it does not give rise to any complication.

IV. CRYSTAL GROWTH

The knowledge of phase relationships, as just described,
can be utilized for the preparation of single crystals.

Getting suitable single crystals of a substance very often
is a critical pre-step of a structure investigation as a whole.
Thus it might be of interest to learn details about the crystal
growth technique used with the alkali metal suboxides, a
technique, which easily can be transferred to quite different
chemical systems. From a more general point of view it means,
how one can use equilibria and non-equilibria occurring with
such binary or higher systems for the preparation of single
crystals.

It can be anticipated that the softness and rather wax-like
consistency of the alkali metal suboxides together with the
pronounced reactivity and low temperatures of decomposition and/or
melting makes handling of single crystals an extremely difficult
task. The only promising way seems to be growing the crystals
inside an x-ray capillary under suitable conditions. This
procedure is known for example from the work of Lipscomb [19] on
some congruently melting molecular crystals. But the method can
be applied to more complicated cases as will be shown.

The case of Rb_6O is such a rather complicated example to
start with (fig. 3). One has to deal with the following situation
as may be recalled: Rb_6O is unstable above -8°C. It should be
formed by decreasing the temperature in a solid-state reaction
between rubidium and the compound Rb_9O_2, at the lower temperature of the double effect. But this reaction is inhibited for
kinetic reasons and, of course, a single crystalline Rb_6O sample
does not result in this way. So large crystals of the phase
Rb_6O can only be grown directly from the melt, which should be
impossible at all under equilibrium conditions and is indeed
impossible in practice with the proper sample composition,
because a lot of crystals (serving as seed crystals) of Rb_9O_2
in the melt favour the equilibrium condition. Now by Guinier
experiments it can easily be shown, that Rb_6O crystallizes
directly from a melt if two conditions are met:

(a) the melt must contain an excess of metal (about 12 at.-% oxygen at most). This leads to a delayed crystallization of the competing phase Rb_9O_2 due to the absence of seed crystals,

(b) the cooling procedure is of great importance and decides which of both phases crystallizes. By quenching from room temperature to $-30°C$, the wanted phase Rb_6O is reproducibly generated.

Thus single crystals are derived as shown in fig. 5:

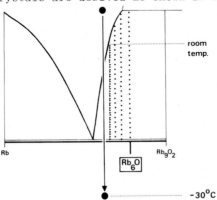

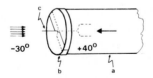

Fig. 5. Single crystal growth of Rb_6O.
(upper part: simplified Rb_2O/Rb phase diagram, the region of primary crystallization of Rb_9O_2 at room temperature indicated by triangles;
lower part: crystal growth device; a = layer line screen of Weissenberg instrument, b = paper screen, c = sample capillary).

A capillary filled with a liquid sample of the rubidium-rich composition is introduced into a Weissenberg instrument with a cooling device; the line screen opposing the stream of cooling gas is closed at the end by two specially cut overlapping sheets of paper in a frame with a little hole left in the center. In the tip of the capillary, reaching into a stream of cold gas, seed crystals of Rb_6O are generated, whereas the content of the capillary behind the paper screen is kept liquid with warm air. By slowly moving the capillary into the cold region, large single

crystals of Rb_6O grow in the steep temperature gradient. The
excess of rubidium present fortunately does not disturb further
investigations. At the end the paper screen is withdrawn.

An example of a very simple case, on the contrary, is given
by the compound Cs_7O, because this one melts congruently. As
shown in fig. 6 a capillary filled with a liquid sample at
room temperature is surrounded by a gas stream, the temperature
of which is kept slightly below the melting point of the sample.
This remains liquid due to supersaturation (see p. 53,54). By
touching the end of the capillary by a wire cooled with liquid
nitrogen, the whole content solidifies from that point resulting
in a cylindrical single crystal, which is perfectly aligned with
one crystallographic axis parallel to the axis of the capillary.

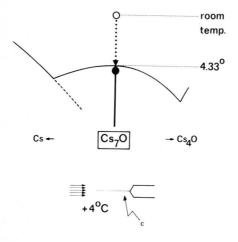

<u>Fig. 6.</u> Single crystal growth of Cs_7O.
(upper part: liquidus of the Cs_4O/Cs phase diagram;
lower part: schematic drawing of the crystal growth
technique; c = sample capillary on goniometer head).

The growth conditions for single crystals of another compound,
$Cs_{11}O_3$, are rather uncomplicated too, mainly because the phase
is stable at room temperature. So crystal growth can take place
outside the camera. As fig. 7 shows, the special phase diagram
principally suggests two different ways of crystal growth:
With the first, based on non-equilibrium conditions (situation I),
the totally liquid sample is rapidly cooled to a temperature below
the decomposition temperature of $Cs_{11}O_3$. Thereby it is tried to

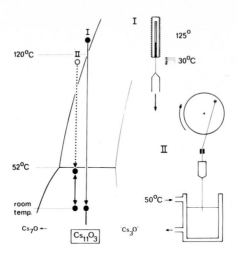

<u>Fig. 7</u>. Single crystal growth of $Cs_{11}O_3$.
(left: simplified region of Cs_2O/Cs phase diagram,
I, II indicate different procedures of crystal growth as
explained in the text;
right: realisations of the crystal growth procedures I
and II).

prevent crystallization of Cs_3O. This can be done by slowly
withdrawing the capillary filled with liquid, from a heated,
narrowly drilled metal block. Care has to be given to attain a
steep temperature gradient at the end of the block by a stream
of cool air blowing from the side. But the crystal growth of
$Cs_{11}O_3$ is strongly disturbed, because the other phase in
competition, Cs_3O, crystallizes very easily. It is therefore
much better to work under equilibrium conditions, which means
at temperatures below the decomposition temperature of $Cs_{11}O_3$
all the time (situation II). The capillary is then filled with
a liquid sample containing an excess of cesium as in the
case of Rb_6O. It is then rapidly cooled to generate polycrystalline $Cs_{11}O_3$, suspended in a cesium-rich melt. By simply
oscillating the sample as shown in fig. 7, thereby dipping it
repeatedly into a bath of $50°C$, a large single crystal of $Cs_{11}O_3$
has been grown in the upper part of the capillary after about
one week. Again the crystal was perfectly aligned with one axis
parallel to the axis of the capillary and well separated from
the excess of melt.

Last not least there is the really difficult problem of

the growth of Cs_4O single crystals. As recapitulated in fig. 8

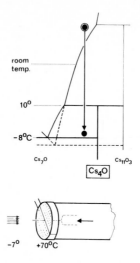

Fig. 8. Single crystal growth of Cs_4O
(explanations as in fig. 5).

this phase decomposes below room temperature and we never have
been able to generate the phase by cooling tiny amounts of
substance in an x-ray capillary, because then the non-equilibrium condition with $Cs_{11}O_3$ as the governing phase is favoured.
Growing single crystals of Cs_4O therefore calls for a rather
complicated procedure.

One starts with some 50 grams of a material with a composition containing an excess of cesium. The bulk is patiently
persuaded to form the equilibrium phase Cs_4O. After this has
happened more or less by chance, and indicated by its thermal
behaviour, it is kept cool for the rest of the time serving
as a reservoir of the stable phase. A sample is extracted now
and, while still liquid, brought into the capillary; thereby,
of course, the Cs_4O is decomposed. But in cooling the melt
inside the capillary, the stable phase is formed again by
adding some cold grains from the main bulk. The capillary thus
contains polycrystalline Cs_4O plus melt. Under permanent cooling
it is closed and transferred to the camera or diffractometer
for the single crystal growth procedure, quite similar to the
one described for Rb_6O. Instead of a paper screen, the capillary
reaches here into the nitrogen stream through a removable metal

screen with inserted heating elements. A long single crystal
of Cs_4O is grown from the melt on the seed crystals at the top
by slowly moving the capillary into the cold region.

V. CRYSTAL STRUCTURES

As a kind of reward for such a lot of necessary experimental
work, we were lucky to find the alkali metal suboxides to be a
very uniform group of compounds with a unique structural principle.

The main principle of the crystal structures of the alkali
metal suboxides is quite simple [8-10], as illustrated schematically in fig. 9: in all cases the oxygen atoms are surrounded

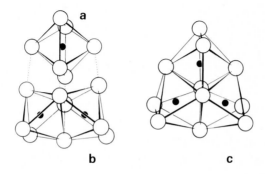

Fig. 9. Coordination polyhedra occurring in alkali metal suboxides (open circles: Rb,Cs; see text).

octahedrally by metal atoms (fig. 9a). Two coordination
octahedra may be face-sharing. This results in a cluster of
atoms with composition M_9O_2 (fig. 9b). A third octahedron can
be added to this cluster resulting in a trigonal group with
the stoichiometric composition: $M_{11}O_3$. Each coordination
octahedron shares two adjacent faces with others (fig. 9c).

Both stoichiometric compositions, M_9O_2 and $M_{11}O_3$, are
occurring for the suboxides of rubidium and cesium. To start
with the last one, fig. 10 shows the crystal structure of
$Cs_{11}O_3$ as projected parallel [010] of the monoclinic unit cell
[8]. It is obvious, that the compound $Cs_{11}O_3$ consists of discrete clusters of the same stoichiometry $Cs_{11}O_3$, which are
well separated from one another. The cluster geometry is
exactly the same as derived in fig. 9 except for some small
deviations from the ideal trigonal symmetry caused by the site

A. Simon

symmetry of the group.

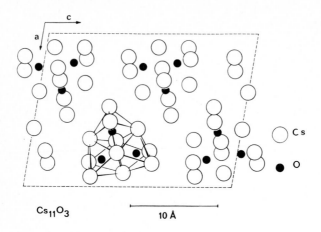

$Cs_{11}O_3$ |———— 10 Å ————|

Fig. 10. Structure of $Cs_{11}O_{10}$.

    Table III contains mean interatomic distances, which hold for
$Cs_{11}O_3$ as well as for the other cesium suboxides discussed later.

Table III

Interatomic distances in $Cs_{11}O_3$ (Å)

within $Cs_{11}O_3$ groups:	between $Cs_{11}O_3$-groups:
Cs-Cs 3.75 - 4.3	minimum 5.3
Cs-O 2.7 - 3.0	
O-O 4.0	

These distances should be compared with those in Cs_2O (anti-
$CdCl_2$) having the same oxygen coordination [20]. Here Cs-Cs
distances are 3.8, 4.2 and 4.3 Å, very similar indeed to the
corresponding values within the $Cs_{11}O_3$-clusters, whereas the
shortest Cs-Cs contacts in metallic cesium itself (5.3 Å)
directly correlate to inter-cluster distances in $Cs_{11}O_3$. The
mean Cs-O distance in $Cs_{11}O_3$ is very similar to the corresponding distance in Cs_2O (2.86 Å) again, which equals the sum of
the ionic radii. But there is a remarkable shift of the oxygen
atoms towards the periphery of the $Cs_{11}O_3$ cluster. It gives the
impression of a strong repulsion between the oxygen atoms with
the consequence of extremely short distances of the oxygen
atoms to the peripheral cesium atoms (2.7 Å) as compared with

the elongated distances to the central cesium atoms (3.0 Å). The
cage of cesium atoms follows this pattern, hence the shared
faces of the coordination octahedra are very much contracted
as compared with the others. The high polarizability of cesium
seems to be an important factor in dealing with these distortions.

The structure of Rb_9O_2 serves as the Rb-analogue of $Cs_{11}O_3$;
the structure [8] is seen projected along [010] of the monoclinic
unit cell in fig. 11.

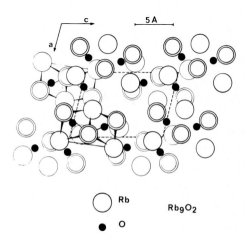

Fig. 11. Structure of Rb_9O_2.

Not quite as evident as in the case of $Cs_{11}O_3$, but this only
because of the angle of projection, the structure is built
of clusters that consist of two face-sharing coordination
octahedra looked upon from the side. Two of these are indicated
by outlining in fig. 11. Again the shortest distances between
such clusters are 5.1 Å - shortest Rb-Rb distances in metallic
rubidium are 4.8 Å - whereas the Rb-Rb-distances within the
clusters (3.5 Å for the atoms belonging to the common face of
both coordination octahedra, the others ranging from 3.8 to
4.0 Å) partly resemble those in the normal ionic oxide Rb_2O
(Rb-Rb-distance of 3.38 Å in γ-Rb_2O [21]). In the Rb_9O_2-group
too the oxygen atoms are shifted to a large O-O-distance of
4.0 Å resulting in remarkably short distances to the outer Rb-

atoms (shortest distances 2.6 Å). Except for this displacement
the Rb_9O_2-cluster entirely corresponds to the $W_2Cl_9{}^{3-}$-cluster
mentioned earlier [14].

So, in the structures of both Rb_9O_2 and $Cs_{11}O_3$ characteristic clusters occur, well separated from one another. Metal-metal
distances and metal-oxygen distances within the clusters do
agree fairly well with those in the normal oxides of the corresponding alkali metals, whereas the interatomic distances
between different clusters are by far longer and very much
comparable to the distances in pure alkali metals.

This fact together with the physical properties of these
compounds - metallic lustre and metallic conductivity [4,7] -
could be explained on the basis of a yet very qualitative
picture [10]: If "normal" oxydation states of oxygen and alkali
metal are assumed (-2 and +1 respectively), the Rb_9O_2- and
the $Cs_{11}O_3$-clusters have to be charged +5, free electrons compensating this charge in the neutral compounds. So at least
part of the excess electrons are delocalized in a conduction
band. Therefore this kind of compounds is best described with
the term "complex metals".

Besides the coordination principle, which to our present
knowledge specifically leads to the clusters of compositions
Rb_9O_2 and $Cs_{11}O_3$, there exists a second structural principle
leading to a variety of different though strictly stoichiometric
compositions of the suboxide phases: Rb_9O_2 and $Cs_{11}O_3$ can take
up certain amounts of pure alkali metal to form new compounds.
Looked upon from another point of view the characteristic
clusters occurring in Rb_9O_2 and $Cs_{11}O_3$ can be suspended in a
stoichiometric amount of pure alkali metal to form a well-ordered
array.

This is illustrated in fig. 12 for the compound Rb_6O.
The crystal structure is projected along [100] of the hexagonal
unit cell (dotted line) [8]. For sake of clearness the identity
period is repeated several times. It is quite obvious from fig.
12, that in the structure of Rb_6O the Rb_9O_2-clusters (one of
them is outlined) as preferred by the element rubidium occur
arranged in layers, and these layers alternate with close packed
planes of rubidium. Some indication of the character of the
intermediate Rb-planes may be obtained from the interatomic
distances. These are 4.87 Å and 4.95 Å within the planes and
4.87, 5.21 and 5.30 Å to the nearest lying atoms of adjacent

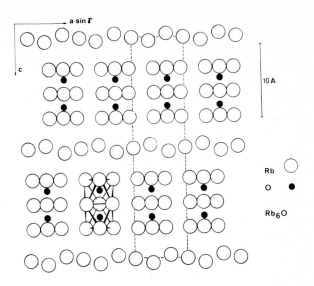

__Fig. 12.__ Structure of Rb_6O; unit cell indicated by dotted lines.

Rb_9O_2-layers. The resulting mean distances of 5.02 Å around each atom of the intermediate Rb-planes agree fairly well with metallic distances, if one takes into account the high coordination number of these atoms in Rb_6O. From a structural point of view, the formula of Rb_6O can thus be written as 2 Rb_6O = $[Rb_9O_2]Rb_3$.

In the still more interesting case of cesium, different compounds $(Cs_7O$ and $Cs_4O)$ exist containing the same clusters as $Cs_{11}O_3$ even in geometrical details, but different in the amount of additional alkali metal atoms. That is shown for the compound Cs_7O in fig. 13. Again the hexagonal unit cell as projected along [001] is repeated several times to work out the main principle [8]. This time the trigonal $Cs_{11}O_3$-clusters are lined up to form piles pointing towards the plane of drawing. The piles are surrounded by purely metallic regions of partly hexagonal close packed cesium. Hence Cs_7O is a one-dimensional analog of Rb_6O. Again from a structural point of view Cs_7O is well described as $[Cs_{11}O_3]Cs_{10}$.

The structures looked at so far follow a very intriguing pattern. They can be discussed in terms of salt-like units held

A. Simon

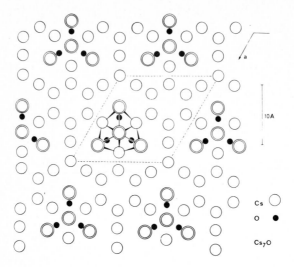

Fig. 13. Structure of Cs_7O; hexagonal unit cell indicated by dotted lines.

together by a metallic type of bonding in the compounds Rb_9O_2 and $Cs_{11}O_3$. The structures of Rb_6O and Cs_7O serve as even more interesting examples of intermediates between salt-like and metallic types of bonding, especially because of the distinct separation between the different bonding regions in space. These regions are stretched throughout the crystal.

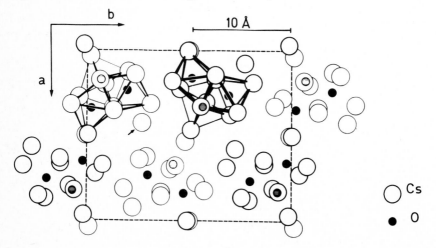

Fig. 14. Structure of Cs_4O; orthorhombic system; single Cs-atom indicated by an arrow.

In the crystal structure of Cs_4O, which we were able to solve in the meantime, the above mentioned pattern shows a very peculiar modification [22]. As is easily concluded from fig. 14, the compound Cs_4O can be described as $[[Cs_{11}O_3]Cs$. This time the "dilution" of $Cs_{11}O_3$ by an excess of alkali metal leads to isolated Cs-atoms distributed in the structure without having direct contacts with one another as they have in the cases of Rb_6O and Cs_7O.

These additional cesium atoms are not coordinated to oxygen in any sense and the shortest distance to another Cs-atom of the neighbouring clusters is 5.3 Å.

It means that we meet the very unusual situation with the alkali metal suboxides, that purely metallic districts occur in the structures of these compounds extending in three dimensions in the case of Cs_7O and in two dimensions in the case of Rb_6O, whereas they are confined to single atoms, well isolated in space, in the case of Cs_4O. Of course, this picture seems less artificial, if the alkali metal suboxides are discussed in terms of some kind of intermetallic compounds formed between the "complex metals" Rb_9O_2 or $Cs_{11}O_3$ and the pure alkali metals.

VI. PROSPECTS

Still there are several unanswered questions as to the chemistry, structure and nature of the alkali metal suboxides.

There might possibly exist more binary compounds, we do not know yet. Kinetic effects as those found with the phases Cs_4O might well hide some existing compound. Moreover the structure of Rb_6O (fig. 12) has a kind of variable pattern and it should be possible to find two, three or even more intermediate metal atom layers between the Rb_9O_2 layers instead of just only one. Indeed there is experimental evidence for the existence of rubidium suboxides with still higher metal contents. But as decomposition and formation of these compositions seem to be confined to the very narrow temperature interval between the Rb/Rb_9O_2-eutectic and the decomposition temperature of Rb_6O, phase investigations have become extremely difficult. This leads us to the next question.

Still there is no true understanding of the possible rules governing the peculiar stoichiometries of the alkali metal suboxides, which, of course, is a two-sided problem. On one hand

to our present knowledge, the types of complex groups found
with alkali metal suboxides seem to be confined to those of
Rb_9O_2 and $Cs_{11}O_3$. This, of course, is a limiting factor to the
stoichiometry. On the other hand, the intermediate parts of the
structures consisting of pure metal are obviously variable to
a certain extent only. The factors that influence the extension
of these parts, whether of electronic nature or due to size
relations, are yet unknown. It is hoped to find some answer
to these questions by investigating mixed Rb/Cs-suboxides.
With respect to the rather great number of binary compounds,
it is not too surprising that an at least equal number of ternary
compounds does exist between -50°C and +50°C. We hope to be
able to extract some stoichiometry rule out of a greater variety
of such compounds.

Last not least, there remains the question whether this
kind of compounds is restricted to the metals rubidium and
cesium only. Of course, one looks for more compounds with
potassium first. According to the K/K_2O-phase diagram [23] in
comparison with the Rb/Rb_2O- and Cs/Cs_2O-diagrams, suboxide
formation would be possible by solid-state reactions only. We
tried to overcome these difficulties by oxidizing liquid
potassium alloys at low temperatures [24,25]. It turned out
that potassium participates in suboxide formation, at least in
forming mixed compounds with cesium. But as these compounds
with cesium need to be prepared and investigated at temperatures below -50°C, their further investigation is a forthcoming
work, asking for even more sophisticated experimental techniques.

The very engaged cooperation of W. Brämer, H.-J. Deiseroth and
E. Westerbeck is gratefully acknowledged.

References

[1] H. Föppl, Z. Anorg. Allg. Chem. 291 (1957) 12.

[2] E. Rengade, Bull. Soc. Chim. France $\begin{bmatrix}4\end{bmatrix}$ 5 (1909) 994.

[3] E. Rengade, Compt. Rend. (Paris) 148 (1909) 1199.

[4] G. Brauer, Z. Anorg. Chem. 225 (1947) 101.

[5] Ph. Touzain, Can. J. Chem. 47 (1969) 2639.

[6] Ph. Touzain and M. Caillet, Rev. Chim. Miner. 8 (1971) 227.

[7] Ph. Touzain, Bull. Soc. Chim. France (1972) 4515.

[8] A. Simon, Naturwissenschaften 58 (1971) 622, 623.

[9] A. Simon and E. Westerbeck, Angew. Chem. 84 (1972) 1190;
Intern. Ed. 11 (1972) 1105.

[10] A. Simon, Jahrb. Akad. Wissensch. Göttingen (1972) 19.

[11] A. Simon, Z. Anorg. Allg. Chem. 395 (1973) 30.

[12] D. Bauer and H.-G. von Schnering, Z. Anorg. Allg. Chem. 361
(1968) 259.

[13] C. Brosset, Arkiv Kemi Mineral. Geol. A20 (1946) 7;
A22 (1947) 11.

[14] W.H. Watson, Jr. and J. Waser, Acta Cryst. 11 (1958) 689.

[15] D. Babel and W. Rüdorff, Naturwissenschaften 51 (1964) 85.

[16] K.-R. Tsai, P.M. Harris and E.N. Lassettre, J. Phys. Chem.
60 (1956) 345.

[17] A. Simon, J. Appl. Cryst. 3 (1970) 11.

[18] A. Simon, J. Appl. Cryst. 4 (1971) 138.

[19] W.J. Dulmage and W.N. Lipscomb, Acta Cryst. 5 (1952) 260.

[20] K.-R. Tsai, P.M. Harris and E.N. Lassettre, J. Phys. Chem.
60 (1956) 338.

[21] Ph. Touzain and M. Caillet, Rev. Chim. Mineral. 8 (1971) 277.

[22] A. Simon, H.-J. Deiseroth and E. Westerbeck, unpublished
(1974).

[23] F. Natola and Ph. Touzain, Can. J. Chem. 48 (1970) 1955.

[24] E. Westerbeck, Diplomarbeit, Münster (1971).

[25] W. Brämer, Diplomarbeit, Münster (1972).

Crystal Structure and Chemical Bonding in Inorganic Chemistry
Eds. C.J.M. Rooymans and A. Rabenau
© 1975, North-Holland Publishing Company, The Netherlands

STRUCTURES AND CHEMICAL BONDS IN VANADIUM BRONZES

Paul Hagenmuller

Laboratoire de Chimie du Solide du C.N.R.S.

Bordeaux-Talence, France

I. INTRODUCTION

Unlike tungsten bronzes and even molybdenum bronzes comprising a small number of relatively simple structures, vanadium bronzes exhibit a very large variety of structural types and even formulae [14]. The $V_2O_5-VO_2-K_2O$ diagram is significant in this connection: we observe phases with the formulae $K_xV_2O_5$, $K_xV_3O_8$ and even $K_{2-x}V_{3+2x}O_{8+2x\pm y}$, the last one corresponding to a two-dimensional existence domain [1] (Fig. 1).

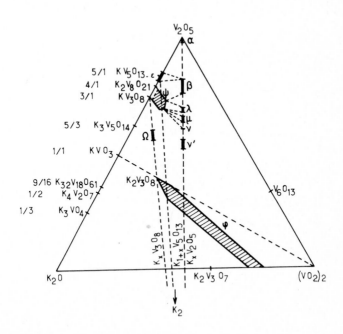

<u>Fig. 1</u>. The $V_2O_5-(VO_2)_2-K_2O$ system at 550°C.

69

The diversity of the structures of vanadium bronzes is due to
the variety of the co-ordination polyhedra of vanadium when its
oxidation state   is between 5+ and 4+: the triangular
bipyramid and the square-base pyramid - more or less distorted -
occur between the tetrahedron and octahedron which may be
regular or irregular. These polyhedra form a covalent lattice,
with a formula $\left(V_2O_5\right)_n$, for example, in the case of $M_xV_2O_5$
bronzes. The M^{p+} ions insert themselves in the oxygen tunnels
of the lattice, releasing inside a chain of similar polyhedra px
electrons which occupy equivalent orbitals, predominantly d,
between which they move by a hopping mechanism. The nature of
the orbitals in which the d-type electrons are localized influences that of the polyhedra concerned. The crystal structure
of the covalent lattice is therefore linked to the conduction
mechanism. Research into these mechanisms is largely based on
a model proposed by Goodenough for $M_xV_2O_5$ bronzes [2].

Two examples will enable us to show how, on the one hand,
an exact knowledge of the structure allows the conduction mode
$\left(\varepsilon\text{-}Cu_xV_2O_5\right)$ to be determined and how, on the other hand, determination of the mechanism on the basis of magnetic ane electrical
properties allows us to decide between two possible structures
$\left(\alpha'\text{-}Na_xV_2O_{5-y}F_y\right)$.

II. ε - $Cu_xV_2O_5$ PHASES

First demonstrated by Casalot [3], these phases contain a
relatively high proportion of copper ($0.85 \leqslant x < 1$ at 620°C). They
are of monoclinic symmetry and belong to the group Cm. Fig. 2
shows the projection of the structure on the [010] plane [4].
The vanadium atoms occupy the centres of greatly distorted
octahedra which are present in groups of four with common edges.
These blocks arrange together to form double layers parallel
to the [001] plane. The copper atoms are positioned between
these layers in two different crystallographic sites, Cu_1 and
Cu_2.

The oxygen environment of Cu_1 is of the octahedral type,
while that of Cu_2 is a parallelogram. For both sites two Cu-O
distances are shorter than the others (Cu_1-O_2' = 2.03 Å and
Cu_1-O_7= 1.99 Å, while $2Cu_1$-O_9'= 2.28 Å and $2Cu_1$-O_4'= 2.50 Å;

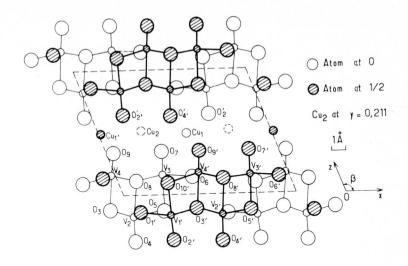

<u>Fig. 2</u>. Projection of the structure of ε-$Cu_xV_2O_5$ on the [010] plane.

Cu_2-O_2' = 1.94 Å and Cu_2-O_7 = 2.02 Å, while Cu_2-O_4' = 2.50 Å and Cu_2-O_9 = 2.42 Å). These short bonds form linear or quasi-linear O-Cu-O chains which might be considered to represent the sp hybridization of the monovalent copper. The presence of a flattened octahedron round Cu_1 and that of a plane quadrilateral round Cu_2 could, however, also indicate a Jahn-Teller effect due to the divalent copper. Only measurement of the physical properties makes it possible to determine the exact oxidation state of copper and also the location of the d electrons on the four distinct vanadium atoms in the blocks that form the lattice [3,5].

The problem of the oxidation state of copper was settled by a magnetic investigation. The χ^{-1} = f(T) curves characterize a Curie-Weiss behaviour above a Néel point which is of the order of 125 K for all values of x (Fig. 3). The values experimentally found of the molar Curie constants are compared in Table 1 with those of the theoretical values corresponding to the configurations:

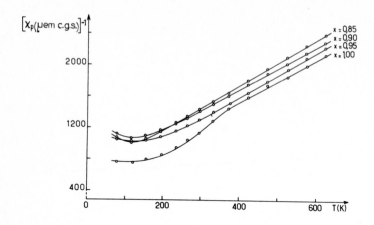

Fig. 3. Variation of the inverse of the magnetic susceptibility
of $\varepsilon\text{-}Cu_xV_2O_5$ phases with temperature.

$$Cu^+_x V^{4+}_x V^{5+}_{2-x} O_5 \qquad (1)$$

and

$$Cu^{2+}_x V^{4+}_{2x} V^{5+}_{2-2x} O_5 \qquad (2)$$

i.e. $0.375 \cdot x$ and $1.125 \cdot x$, on the assumption that the only contribution is due to spin. The factor α denotes the fraction
of the copper that is theoretically in the monovalent state.
Table 1 shows that in the neighbourhood of $x = 1$ the copper is
essentially monovalent. The constancy of the Néel temperature

TABLE 1

x	C_{mol} (exp.)	C_{mol} (theor.) (1)	C_{mol} (theor.) (2)	α
0.85	0.342	0.318	0.954	0.96
0.90	0.366	0.338	1.014	0.96
0.95	0.355	0.356	1.068	1.00
1.00	0.377	0.375	1.125	1.00

also tends to confirm the relatively small variation in the
number of V^{4+} involved at $T < 125$ K in the anti-ferromagnetic interactions for $0.85 \leqslant x < 1$, as yielded by the assumption (1).

The electric conductivity varies exponentially with temperature.
The activation energy is of the order of 0.12 eV for all
values of x. It increases by about 0.02 eV below the Néel point.
The Seebeck coefficient at 300°K changes very rapidly with x
(Fig. 4). It starts off low and negative (α = -10 μ V/K for

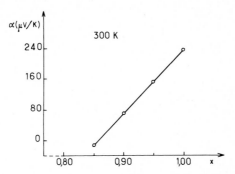

Fig. 4. Variation of the Seebeck coefficient with x for T = 300°K.

x = 0.85), then changes its sign and becomes 225 μ V/K for
x = 1. The positive sign for x = 0.90 characterizes hole con-
duction, while for $0.85 \leqslant x \leqslant 0.90$ the negative sign seems to imply
the presence of p and n carriers simultaneously. The constancy
of α with temperature over the measured range (150 K<T<800 K)
suggests that the number of carriers is independent of the
temperature (Fig. 5). The electrical measurements therefore

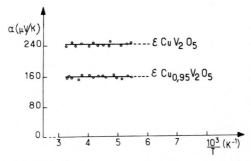

Fig. 5. Variation of the Seebeck coefficient with the reciprocal
temperature.

indicate a hopping mechanism in which the carriers are essentially
p-type, at least in the neighbourhood of x = 1.

The conduction mechanism was investigated with the aid of
the schematic representation in Fig. 6. The V-O distances for
x = 0.85 are given in Table 2.

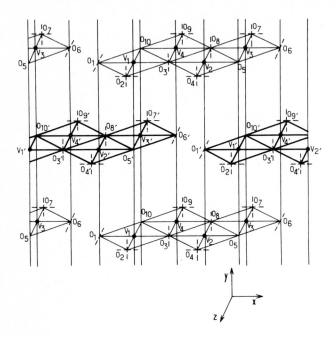

<u>Fig. 6</u>. Schematic representation of the idealized octahedron
chains parallel to the OY axis. The X-axis corresponds
essentially to OX; Z is the third axis of the Cartesian
reference trihedron. The $\sigma(V-O)$ bonds are parallel to
the axes. The anionic p orbitals which are not active in
bonds are represented by dotted lines; they are necessarily involved in $\pi(V-O)$ bonds with the t_{2g} orbitals of the
vanadium.

TABLE 2

Interatomic distances (Å) in $\varepsilon\text{-Cu}_{0.85}\text{V}_2\text{O}_5$

$V_1 - O_1 = 2.01$	$V_2 - O_{1'} = 1.93$
$V_1 - O_2 = 1.62$	$V_2 - O_3 = 1.74$
$V_1 - O_3 = 1.95$	$V_2 - O_4 = 1.60$
$V_1 - O_{5'} = 1.96$	$V_2 - O_5 = 1.99$
$V_1 - O_{10} = 2.28$	$V_2 - O_8 = 2.35$
$V_3 - O_5 = 2.24$	$V_4 - O_3 = 2.27$
$V_3 - O_6 = 1.89$	$V_4 - O_{6'} = 1.89$
$V_3 - O_7 = 1.67$	$V_4 - O_8 = 1.78$
$V_3 - O_8 = 1.87$	$V_4 - O_9 = 1.62$
$V_3 - O_{10'} = 1.92$	$V_4 - O_{10} = 2.16$
$Cu_1 - O_{2'} = 2.03$	$Cu_2 - O_{2''} = 1.94$
$Cu_1 - O_{4'} = 2.50$	$Cu_2 - O_{4''} = 2.50$
$Cu_1 - O_7 = 1.99$	$Cu_2 - O_7 = 2.02$
$Cu_1 - O_9 = 2.28$	$Cu_2 - O_9 = 2.42$

The displacement of the vanadium atoms from the centre of symmetry of their respective octahedra is due to the action of two co-operating forces: cation-cation electrostatic repulsion and cation-anion interactions, the scale of which increases markedly with π-bonding contribution. The latter stabilizes the anionic p orbitals concerned but destablilizes the d orbitals. The d electrons will consequently occupy the t_{2g} orbitals which are not involved in the π-bonding. Analysis of the V-O distances enables us to determine the corresponding bonding index. For vanadium bronzes Goodenough [2] recognizes the following criteria:

$$V\text{-O} < 1.65 \text{ Å} : \text{triple bond,}$$
$$1.65 \text{ Å} < V\text{-O} < 1.85 \text{ Å} : \text{double bond,}$$
$$V\text{-O} > 1.85 \text{ Å} : \text{single bond.}$$

Within the $\varepsilon\text{-Cu}_x\text{V}_2\text{O}_5$ phases all V-O distances correspond to single bonds with the exception of six which are shown together with their multiplicity in Table 3, where the type of d orbitals destabilized by the π-bonding is also indicated.

TABLE 3

Bond	Multiplicity	Orbitals destabilized	Vanadium involved
$V_2 - O_4$	3	d_{yz} and d_{zx}	V_2
$V_1 - O_2$	3	d_{yz} and d_{zx}	V_1
$V_4 - O_9$	3	d_{yz} and d_{zx}	V_4
$V_3 - O_7$	3	d_{yz} and d_{zx}	V_3
$V_2 - O_3$	2	d_{xy}	V_2
$V_4 - O_8$	2	d_{xy}	V_4

The Table shows that the only orbitals which are non-bonding
and therefore liable to accomodate d electrons are the d_{xy} orbitals
of V_1 and V_3. At most, therefore, two d electrons will be
located in the four possible vanadium sites. This situation
corresponds to the top value of x $(x = 1)$ and to the formula:

$$Cu^+V^{4+}V^{5+}O_5$$

which was in fact observed experimentally.

On the other hand, since $V_1-O_5' = 1.96$ Å and $V_1-O_1 = 2.01$ Å,
while $V_3-O_{10'} = 1.92$ Å and $V_3-O_6 = 1.89$ Å, it is clear that
the d_{xy} orbital of V_1 is more stable than the d_{xy} orbital of V_3,
being therefore preferentially occupied by d electrons. The hole
conduction indicated by the transport properties will take place
essentially along the $V_3-O_{10'}-V_3$ chains parallel to the Y-axis
via the d_{xy} orbitals.

III. $\alpha'-Na_xV_2O_{5-y}F_y$ PHASES

The $\alpha'-Na_xV_2O_5$ phases were observed by Pouchard et al.
[1,6]: at 600°C $0.70 \leqslant x \leqslant 1$. The oxygen can be partially
replaced by fluorine in accordance with the coupled substitution:

$$V^{5+} + O^{2-} = V^{4+} + F^-,$$

resulting in a two-dimensional $\alpha'-Na_xV_2O_{5-y}F_y$ domain within
the $V_2O_5 - NaV_2O_5 - NaV_2O_4F$ diagram. The width of this domain
has been determined by Galy and Carpy [7] (Fig. 7).

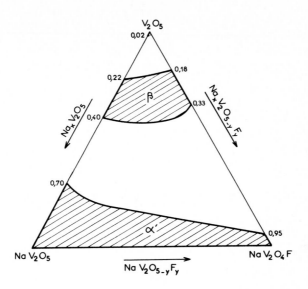

<u>Fig. 7</u>. The V_2O_5 - NaV_2O_5 - NaV_2O_4F system at 550°C.

The NaV_2O_5 $(x = 1, y = 0)$ and NaV_2O_4F $(x = 1, y = 1)$ phases
have crystal parameters which are very close to each other
(Table 4), yet their structures differ: in the unit cell of
NaV_2O_5 the vanadium occupies two different sites V_1 and V_2
(it may be thought that the former site, being larger,
preferentially contains 4+ vanadium and the latter 5+ vanadium);
the lattice of NaV_2O_4F differs from that of NaV_2O_5 by the
existence of a mirror plane perpendicular to the [100] axis
and the entirely tetravalent vanadium occupies only one type

<u>TABLE 4</u>

Comparison of crystallographic data

	NaV_2O_5	NaV_2O_4F
Lattice constants	a = 11.318 ± 0.005 Å	a = 11.318 ± 0.005 Å
	b = 3.611 ± 0.002 Å	b = 3.609 ± 0.002 Å
	c = 4.797 ± 0.003 Å	c = 4.802 ± 0.003 Å
Existence conditions	hk0: h + k = 2n	hk0: h + k = 2n
Space group	C_{2v}^7, $P2_1mn$	D_{2h}^{13}, Pmmn
Density (exp.)	3.42 ± 0.02	3.50 ± 0.04
Density (calc.)	3.47	3.52
Z	2	2

óf site [8,9]. In NaV_2O_5, in fact, the vanadium at V_1 and V_2
are surrounded by bipyramids with triangular bases, while
in NaV_2O_4F the co-ordination polyhedron is nearly a square-based
pyramid. The two structures are compared in Fig. 8.

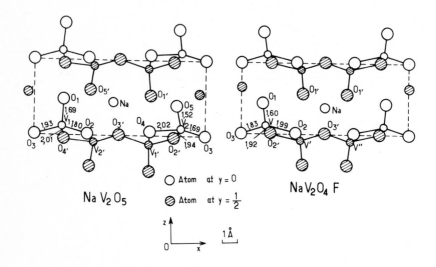

Fig. 8. Projection of the lattice of NaV_2O_5 and NaV_2O_4F on the
[010] plane (distances in Å).

Between the phases NaV_2O_5 and NaV_2O_4F there is a continuous
solid solution $NaV_2O_{5-y}F_y$. The closeness of the lattice
constants for the limiting phases and the impossibility of
preparing a single crystal for $0<y<1$ prevented us from
determining whether the structural evolution was continuous
or whether it occurred suddenly in the neighbourhood of
$y = 0$ or $y = 1$. It was the study of the physical properties
that enabled us to solve this problem [10]. A schematic
representation derived from Pouchard [1] shows that in fact
two structures are possible. In Fig. 9 the $(V_2O_5)_n$ lattice
of NaV_2O_5 is drawn schematically as double chains of octahedra
with common edges parallel to the Oy axis and linked by the
available vertices. Each double chain results from the
association of two single chains of which one contains the 4+
vanadium (A) and the other the 5+ vanadium ions (B).

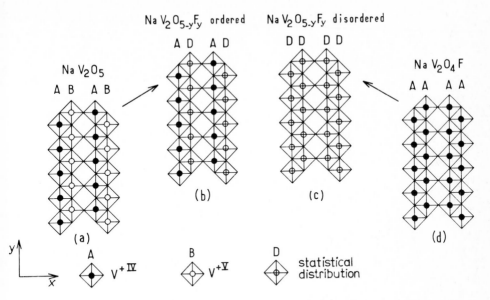

Fig. 9. Development of $(V_2O_5)_n$ chains along the OY axis in the idealized lattice of the $NaV_2O_{5-y}F_y$ phases.

The lattice of NaV_2O_4F can be schematized in the same way but in that case all the chains are of type A. The change from the NaV_2O_5 structure to the NaV_2O_4F structure can occur in two ways: either the excess 4+ vanadium may statistically occupy the B chains (A + D structure) or the whole of the 4+ vanadium may be divided between the two chains (D + D structure). In the former case the crystal structure of NaV_2O_5 persists but evolves progressively towards that of NaV_2O_5F; in the latter case, the NaV_2O_4F structure appears straight off.

Physical investigations show that it is the second hypothesis which is correct. Fig. 10 demonstrates that the Néel point, which is 320 K for NaV_2O_5, decreases rapidly to 150 K when fluorine is introduced , then remains constant to y = 1.

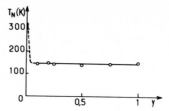

Fig. 10. Variation of the Néel point with y in $NaV_2O_{5-y}F_y$.

Fig. 11 similarly indicates that the activation energy above the Néel temperature, measured from $\log\sigma = f(1/T)$ curves, is 0.12 eV for NaV_2O_5, increases suddenly when the fluorine appears, then remains constant up to the NaV_2O_4F composition.

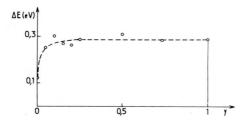

Fig. 11. Variation of the activation energy of the $NaV_2O_{5-y}F_y$ phases with y.

Investigation of the conduction mechanism in the NaV_2O_5 and NaV_2O_4F phases, based on an analysis of the V-O distances, fully confirms that the structure of $NaV_2O_{5-y}F_y$ is similar to that of NaV_2O_4F. Fig. 12 is a schematic representation of the structures of NaV_2O_5 and NaV_2O_4F in which the vanadium co-ordination polyhedra are idealized in the form of elongated octahedra along the Oy axis (the sixth oxygen is located more than 3 Å from V_1, V_2 and V). As Table 5 shows, all V-O distances less than 3 Å correspond to simple bonds except four for NaV_2O_5 and one for NaV_2O_4F, all shown in Table 6, which also lists the t_{2g} orbitals involved in the π-bonding. We assumed that the V-O distance, which is 1.83 ± 0.03 Å,

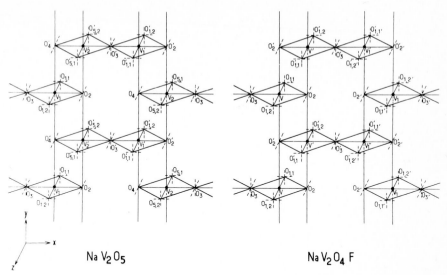

$Na\,V_2\,O_5$ $Na\,V_2\,O_4\,F$

Fig. 12. Schematic representation of the chains of the idealized
octahedra parallel to the Oy axis in the structures of
NaV_2O_5 and NaV_2O_4F. The anionic p orbitals which are
not active in σ bonds are shown by dotted lines.

TABLE 5

Vanadium-oxygen distances ($\pm$ 0.05 Å)

NaV_2O_5		NaV_2O_4F
$V_1 - O_1 = 1.69$	$V_2 - O_3 = 1.69$	$V - O_1 = 1.60$
$V_1 - O_2 = 1.80$	$V_2 - O_4 = 2.02$	$V - O_2 = 1.99$
$V_1 - O_3 = 1.93$	$V_2 - O_5 = 1.52$	$V - O_3 = 1.83$
$V_1 - O_{4'} = 2.01$	$V_2 - O_{2'} = 1.94$	$V - O_{2'} = 1.92$
$V_{1'} - O_{1'} = 3.11$	$V_{2'} - O_{5'} = 3.31$	$V' - O_{1'} = 3.22$

corresponded to a single bond. Table 6 shows that in NaV_2O_5
only the d_{xy} orbital of V_1 is non-bonding and therefore able
to contain a d electron - a result which is compatible with
the limiting formula $NaV^{4+}V^{5+}O_5$ for the solid solution $Na_xV_2O_5$.
In the lattice of NaV_2O_4F the d_{xy} orbital is available for the
d electron of the 4+ vanadium.

TABLE 6

Multiple vanadium-oxygen bonds

	Bond	Multiplicity	Orbital destabilized	Vanadium involved
NaV_2O_5	$V_2 - O_{5,1}$	3	d_{yz} and d_{zx}	V_2
	$V_1 - O_{1,1}$	2	d_{yz} or d_{zx}	V_1
	$V_2 - O_3$	2	d_{xy} or d_{zx}	V_2
	$V_1 - O_2$	2	d_{zx}	V_1
NaV_2O_4F	$V - O_{1,1}$	3	d_{yz} and d_{zx}	V

In the lattice of the $Na_xV_2O_5$ phases conduction is due to a hopping mechanism employing holes in the A chains, i.e. between V_1 vanadiums. This result is confirmed by measurement of the thermo-electric power, being positive and practically independent of temperature (Fig. 13).

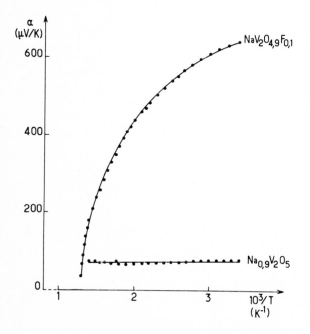

Fig. 13. Thermo-electric power of $Na_{0.9}V_2O_5$ and $NaV_2O_{4.9}F_{0.1}$.

q If the y extra electrons in the $NaV_2O_{5-y}F_y$ phases occupied
the B chains, they would be anti-bonding. On the other hand,
the statistical distribution of the $(1 + y)$ d-electrons in the D
chains only entails the appearance of holes in the non-bonding
d_{xy} levels of NaV_2O_4F. Moreover, in the former case the
conduction, at least for low values of y, would be n-type, while
in the latter case the carriers are obviously p-type. Seebeck
effect measurements confirm that the mechanism involved is
hole conduction. The structure of $NaV_2O_{5-y}F_y$ is thus indeed that
of NaV_2O_4F.

IV. CONCLUSION

The example of the vanadium bronzes illustrates the close
relationships that exist even in relatively complicated solids
between the crystal structure and the magnetic and electrical
properties.

This kind of studies was recently extended to the actual
oxides of vanadium. Thus Casalot, in analysing the location
of d electrons in the cells of V_3O_7 and V_4O_9 prepared by hydro-
thermal synthesis, has confirmed the V_3O_7 formula, but suggested
for V_4O_9 a formula close to $V_4O_8(OH)$ [11] in correspondence
with the structure found by X-ray diffraction by Wilhelmi and
Waltersson [12]. The study of the magnetic properties of V_3O_7
by Bayard, Grenier, Pouchard and Hagenmuller [13], reveals a
spontaneous magnetization of about 1 μ_B at 0°K, confirming
the electronic distribution suggested by Casalot.

References

[1] M. Pouchard, Thèse Doctorat ès sciences, Univ. Bordeaux
    (1967);
    M. Pouchard, J. Galy, L. Rabardel and P. Hagenmuller,
    Compt. Rend. (Paris) 264 (1967) 1943.

[2] J.B. Goodenough, J. Solid State Chem. 1 (1970) 349.

[3] A. Casalot, Thèse Doctorat ès sciences, Univ. Bordeaux (1968).

[4] J. Galy, D. Lavaud, A. Casalot and P. Hagenmuller, J. Solid
    State Chem. 2 (1970) 531.

[5] A. Casalot, D. Lavaud, J. Galy and P. Hagenmuller, J. Solid
    State Chem. 2 (1970) 544.

[6] M. Pouchard, A. Casalot, J. Galy and P. Hagenmuller, Bull.
    Soc. Chim. France 11 (1967) 4343.

[7] J. Galy and A. Carpy, Compt. Rend. (Paris) 268 (1969) 2195.

[8] J. Galy, A. Casalot, M. Pouchard and P. Hagenmuller, Compt.
    Rend. (Paris) 262 (1966) 1055.

[9] A. Carpy and J. Galy, Bull. Soc. Franç. Minéral. Crist. 94
    (1971) 24.

[10] A. Carpy, A. Casalot, M. Pouchard, J. Galy and P. Hagenmuller,
    J. Solid State Chem. 5 (1972) 229.

[11] A. Casalot, Mater. Res. Bull. 7 (1972) 903.

[12] K.A. Wilhelmi and K. Waltersson, Acta Chem. Scand. 24 (1970) 3409.

[13] M. Bayard, J.C.Grenier, M. Pouchard and P. Hagenmuller,
    Mater. Res. Bull., 9 (1974) 1137.

[14] P. Hagenmuller, in: Progress in Solid State Chemistry, Vol. 5
    (Pergamon, Oxford 1971) p. 71.

Crystal Structure and Chemical Bonding in Inorganic Chemistry
Eds. C.J.M. Rooymans and A. Rabenau
© 1975, North-Holland Publishing Company, The Netherlands

STOICHIOMETRY, STRUCTURE AND DISORDER IN SOLID IONIC CONDUCTORS[*]

W.L. Roth

Electrochemistry Branch

General Electric Corporate Research and Development Center
12301 Schenectady, N.Y., U.S.A.

I. INTRODUCTION

Crystal structure and chemical bonding theory have been
extremely successful in accounting for many physical and
chemical properties of solids, particularly, thermodynamic
properties which are determined by the average structure.
Transport phenomena also depend on structure and there is a
current interest in determining the factors which control conductivity and diffusion of ions in solids [1]. Ion transport
depends on the presence of defects or lattice imperfections in
crystals and is accomplished by interchanging atoms between
normal and defect sites. Transport in solids can be described
as the migration of thermally activated lattice imperfections
and the agreement between theory and experiment is satisfactory
when the concentration of defects is small and inter-defect
interactions can be neglected or treated as small perturbations.

The structures and crystal-chemical factors which determine
conductivity are less well understood when the concentration of
defects is large. Chemical formulas frequently do not follow the
law of definite proportions and some solids are highly non-
stoichiometric. Selected area electron diffraction and ultra-
high resolution electron microscopy have shown that the defect
structures in such solids may be exceedingly complex. Interaction
between defects results in the formation of microscopic clusters
or sub-structures in which there are subtle variations in composition and structure [2]. Progress is being made in resolving
the principal morphological features of complex defects but
their precise interatomic arrangements are generally not known.

Parallel with the advances in understanding structures of
non-stoichiometric compounds, a small number of solids have been
discovered which conduct ions at extraordinary high levels at
temperatures well below their melting points. Examples are the

[*] This research was supported in part of AFOSR Contract
F44620-72-C-0007.

silver halides and their chalcogenides, beta-alumina, and several
mixed oxides with the calcium fluoride arrangement. Structure
and conductivity studies have shown anomalous fast transport
through these solids which involves mechanisms differing from
those in conventional materials.

The term "*superionic conductor*" has been used for these
highly conducting solids [3,4]. Based on an analysis of
structure and transport parameters in a phenomenological "free-
ion" theory of ionic transport, solid conductors have been
classified into Types I, II, and III. Type I are conventional
conductors while Types II and III are superionic conductors. The
object of this paper is to describe recent progress in determining the structural factors which control fast ion transport
in two representative superionic conductors: beta-alumina and
calcia-stabilized zirconia.

II. TYPE III CONDUCTORS; BETA - ALUMINA

Type III is the simplest and, consequently, best understood
of the two classes of superionic conductors. Examples are α-AgI,
Ag_4RbI_5, and beta-alumina. A unique aspect of the latter is that
the charge carriers migrate in a single set of widely separated
planes in the crystal. The two dimensional character simplifies
the analysis of the interaction of mobile ions with neighboring
ions and considerable progress has been made in establishing
the mechanism of ion transport and in accounting for crystal-
chemical factors which inhibit or promote conductivity.

Beta-alumina (β) is actually a sodium aluminate in which
the current is carried by Na^+ ions migrating through the crystal
[5]. The conductivity is less than 10^{-2} ohm^{-1} cm^{-1} at room
temperature and increases with temperature. The unique characteristic of ion conduction in β is the small activation energy
or free-ion band gap of only 3.8 kcal mole^{-1}. The free-ion parameters are similar to those for ionic liquids and β can thus be
regarded as a liquid-like solid electrolyte.

II.1 Structure of the Ideal Phases
Beta-alumina is a term that refers to a family of sodium
aluminates with closely related structures and properties, the
most important being hexagonal β and rhombohedral β".

The structure of β was solved by Beevers and Ross in 1937
[6]. The structure has been refined with modern x-ray methods and

the interatomic distances, angles, and thermal parameters are now
accurately known [3,7]. The compound crystallizes in a hexagonal
unit which contains two spinel-like blocks joined by a mirror
plane (Fig. 1). The spinel blocks are composed of four layers
of close packed oxygen atoms with aluminium atoms in tetrahedral
and octahedral interstices. The spinel blocks are bound together
by oxygen columns in the mirror plane. The sodium and oxygen
atoms are loosely packed in a way to allow easy migration of
sodium ions.

     The structure of β" is also well established [8,9,10]. The

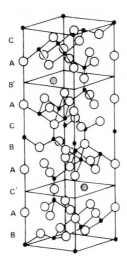

Fig. 1. Structure of β-alumina (idealized). Large open circles,
        oxygen; small solid circles, aluminium; striped circles,
        sodium. The four ABCA layers comprise a spinel block.
        B' and C' are the conduction planes. The Beevers-Ross
        sites are shown completely occupied, the anti B-R sites
        along the cell edge are vacant.

β and   β" arrangements are similar and can be regarded as
hexagonal (2 block) and cubic (3 block) sequences of spinel blocks.
An important difference is that the symmetry of β" is rhombohedral
and neighboring spinel blocks are rotated   π/3 relative to their
configuration in   β. As a result, sodium ions in the conduction
planes of β and   β" occupy sites with different crystallographic
environments and encounter different barriers to motion within
the plane.

     There are numerous beta-like phases with related structures.
The a_o lattice parameters are virtually identical and the c_o

parameters are in the ratio of small integers (Table I). The
phases are the result of different stacking sequences of spinel
blocks. They are pseudo-polytypes with slightly different
stoichiometry and may contain foreign ions. Additives of some
monovalent and divalent cations favor particular sequences; for
example, β' is stabilized by Li^+ and Mg^{2+} [11]. The free
energies of the phases are nearly equal and they easily grow as
mixed crystals [12,13].

TABLE 1

Dimensions of the spinel block in representative isomorphs and
pseudo-polytypes of β-alumina. L is the number of oxygen layers
in a spinel block and N is the number of spinel blocks in a unit
cell.

	L	N	a_o, Å	c_o/N, Å
Li-β	4	2	5.593	11.32
Na-β	4	2	5.594	11.26
K-β	4	2	5.596	11.36
Ag-β	4	2	5.594.	11.25
Na-β"	4	3	5.623	11.20
Na-β"'	6	2	5.62	11.34 x 7/5
Na-β""	6	3	5.62	11.34 x 7/5

Sodium can be replaced by many monovalent cations with only
small changes in the unit cell dimensions [5]. In particular,
the dimensions are nearly independent of the orientation of
neighboring blocks (i.e., whether the phase is hexagonal β or
rhombohedral β") or of the chemical composition in the conduction
planes.

II.2.Defects and Non-Stoichiometry

     The intrinsic defects of particular importance for under-
standing ion transport originate from disorder related to the
non-stoichiometric nature of the compound. In addition, many
defects result from the weak binding between spinel blocks:
crystals are readily cleaved perpendicular to the c-axis and
impurities are easily absorbed into the basal planes. Foreign
ions are frequently added to increase the conductivity of β or
to stabilize a particular structure. Recent neutron diffraction
and nuclear magnetic resonance studies have shown that sub-
stitutions in the spinel block, as well as incorporation of ions

in the conduction planes, can have large effects on the ionic
conductivity.

The crystal and transport properties of single phase β do
not depend strongly on the stoichiometry [14]. At 1725°C, the
phase extends from Na_2O/Al_2O_3 = 1/9.2 to 1/7.8. The unit cell
dimensions and conductivity over this range were found to increase only slightly with the sodium content, consonant with an
increased concentration of sodium in the conduction planes. The
refinement of the site occupation parameters in β indicates that
the charge balance is established by aluminium vacancies in the
spinel block [7]. However, the x-ray density indicates there are
two $Na_{1.22}Al_{11.02}O_{17.14}$ molecules per unit cell, suggesting
the charge of excess sodium is compensated by interstitial oxygen
[3].

Recent neutron diffraction studies show that aluminium
vacancies are probably present in the spinel block but they are
compensated by aluminium atoms in interstitial sites [15].

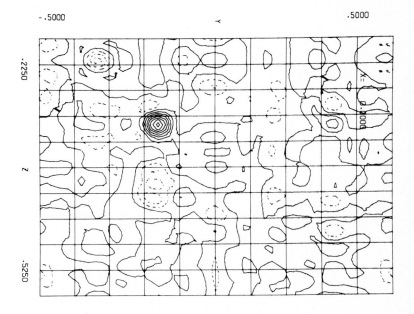

Fig. 2. Nuclear density difference Fourier which shows inter-
        stitial aluminium atom in the spinel block of β-alumina.
        The density corresponds to 0.03 aluminium atoms. (Courtesy
        F. Reidinger)

Fourier maps of nuclear density show that there is a small concentration of nuclear density at an interstitial site in the
spinel block which is adjacent to an aluminium site that contains
a comparable number of vacancies (Fig. 2). These are believed to
be interstitial-vacancy pairs, i.e. *Frenkel defects* . During the
formation of β at high temperature, some aluminium atoms are
probably excited into interstitial positions, leaving vacancies
in the regular lattice sites.

II.3.Stabilization of β" Structure

Foreign ions are frequently added to increase the conductivity of β or to stabilize the β" structure. The conductivity
of β" is greater than β and there is considerable interest in
learning the crystal-chemical reasons for these effects. The
positions of Mg atoms and the mechanism of stabilization of β"

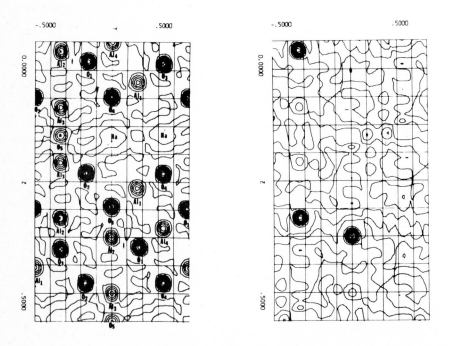

Fig. 3. Nuclear density in ($\bar{1}$10) plane of β"-alumina (left) and
        difference Fourier (right) which shows excess density due
        to magnesium in Al(2) sites of the spinel block.

have been investigated by single crystal neutron diffraction [10].
Neutrons were used because the nuclear form factors are favorable
for distinguishing between Al and Mg atoms. The analysis shows
that Mg replaces Al in one of the four sets of crystallographically independent cation - sites in the spinel block (Fig. 3).
The substituted site is tetrahedrally coordinated and is
characterized in pure β by short Al-O distances. The relevant
interatomic distances in β and   β " are compared in Table 2.
Substitution of Mg in the Al(2) site is accompanied by an increase of 3% in the Al(2)-O distance.

TABLE 2

Bond	Coordination	$R_{Al^{3+}} + R_{O^{2-}}$	β	β"	Δ(%)
Al (1)	6	1.93	1.916	1.914	-0.10
Al (4)	6	1.93	1.895	1.894	-0.05
Al (2)	4	1.79	1.803	1.857	3.00
Al (3)	4	1.79	1.745	1.756	0.63

This finding suggests that there may be localized regions of
high strain due to the small, highly charged Al^{3+} ions in the
Al(2) sites. The strain is reduced with a concomitant stabilization by replacing Al^{3+} with Mg^{2+}, which is both larger and
less positively charged. There is negligible substitution of Al^{3+}
by Mg^{2+} in the other tetrahedrally coordinate site, Al(3),
because the strain can be reduced by bending of the Al(3)-O(5)
bond. The thermal vibration parameters of O(5) in the conduction
plane are extremely large (greater than 0.3 Å) and strain is
reduced by relaxing the O(2)-Al(3)-O(5) bond angle (see Fig. 7).

Another consequence of the smaller strain in the spinel
block of β" is that electric neutrality appears to be established
with relatively few aluminium vacancies or oxygen interstitials
(Table 3).   β" has fewer charge compensating defects than β,
consistent with the fact that it is the equilibrium phase at
low temperature.

TABLE 3

Distribution of Cations in β" Determined by Neutron Diffraction

Atom	Site	Fraction Mg or Na in Site	Number of Atoms per Block Al	Mg	Na
Al (1)	18 h	0.04 ± 0.01	5.8	0.2	
Al (2)	6 c	0.40 ± 0.02	1.2	0.8	
Al (3)	6 c	0.06 ± 0.03	1.9	0.1	
Al (4)	3 a	0.02 ± 0.04	1.0	0	
Na (1)	6 c	0.24 ± 0.04			0.5
Na (2)	18 h	0.19 ± 0.01			1.1
			9.8	1.1	1.6

II.4. Disorder in the Conduction Plane

The conduction plane in β is a state of extreme disorder. The sodium atoms are distributed among two or more sites. The sites are only partially occupied and there is rapid exchange of atoms between them. A considerable amount of information has been obtained about both positional and motional disorder by combining neutron diffraction studies with observations of sodium motion by wide-line nuclear magnetic resonance.

The structural features of disorder have been obtained by least square and Fourier analysis of single crystal diffraction data. Although there are differences in detail, both β and β" are characterized by a statistical distribution of the charge-carrying cations among a large number of similar sites in the conduction plane. In addition, the thermal parameters are very anisotropic and the ions vibrate with exceedingly large amplitudes in the conduction plane.

The distribution of charge carriers among the cation sites was determined in Ag-β to take advantage of the high x-ray form factor of Ag [3]. At room temperature, the Ag was found principally to occupy two sites: the so-called (Beevers-Ross) B-R sites which were 70% occupied and the anti B-R sites which were about 50% occupied (Fig. 4). There is continuous scattering density between the sites due primarily to the large anharmonic vibrations of the Ag ions, but as will be discussed in II.5, some of the density may be due to interstitial oxygen atoms. The distribution found in Ag-β is somewhat different from that reported for Na-β[7]. To determine the relative potential energy

Fig. 4. Electron density in the conducting plane of Ag-β. The
B-R and anti B-R sites are about 2/3 and 1/2 occupied,
respectively. The density in the bridge between sites is
mostly due to silver with a small contribution from
interstitial oxygen.

of the sites in the conduction plane, the temperature dependence
of the distribution of sodium ions in β is being measured by
neutron diffraction (Reidinger, private communication). At room
temperature, most of the sodium atoms are in B-R, anti B-R and
bridge positions; at higher temperature, there is an increased
delocalization of sodium and above 350°C, the disorder is
excessive and the sodium distribution resembles that of a two-
dimensional liquid (Fig. 5); at 180°C, the anti B-R sites are empty.

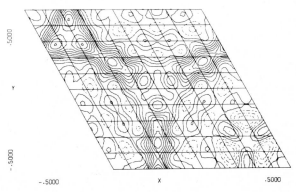

Fig. 5. Distribution of sodium in β -alumina at 600°C from neutron
diffraction. (xy plane). There is a liquid-like distribution of sodium in a hexagonal network of interconnected tunnels joining B-R and anti B-R sites (Courtesy F. Reidinger).

II.5. Enhancement and Inhibition of Sodium Motion

The motion and interaction of sodium with the surrounding crystalline field have been studied by observation of the ^{23}Na spectrum with wide-line NMR (Fig. 6) [16]. The activation energy

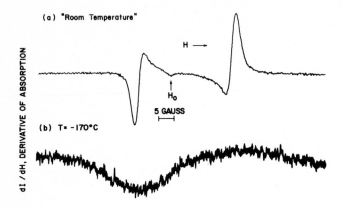

Fig. 6. NMR spectrum of ^{23}Na in polycrystalline β-alumina at room temperature and at -170°C. The room temperature line width is narrowed because of the motion of sodium nuclei (Courtesy I. Chung).

of the sodium motion estimated from the temperature dependent resonance line width agrees with that obtained by diffusion and conductivity (3.9 kcal mole^{-1}). Rotation spectra measured on single crystals show that there is a fast sodium motion among the sodium positions which results in a single symmetric resonance absorption line. The temperature dependence of the quadrupole coupling constant and the asymmetry pattern show that the average position of the sodium ions moves away from the three-fold symmetry axis as the temperature decreases.

The NMR studies have been especially valuable for studying the effects of foreign ions. Measurements of the NMR spectrum of ^{23}Na in β have shown that the mobility of sodium is increased by some additives and decreased by others. The effects of the additives depend primarily on whether they are located in the spinel block or in the conduction planes. Mg, Ni, and Li are examples of the first group, H_2O and Ca are examples of the second group.

   Absorption of water vapor from the atmosphere decreases the
motional narrowing of the ^{23}Na spectrum. The dependence of the
quadrupole coupling constant on water content shows that water
molecules are absorbed in the conducting planes and act as an
effective dielectric by interposing themselves between the
sodium nuclei and the surrounding charges [17]. The ^{23}Na spectrum
has been measured in β specimens that were doped with various
concentrations of Mg, Ni, Li, Y, and Ca (Table 4). The motion of
sodium nuclei is decreased by Ca, not affected by Y, and strongly
enhanced by Mg, Ni and Li. The inhibition by Ca is evidence that
Ca^{2+} ions are incorporated in the conduction plane to form
localized regions which have the $CaAl_{12}O_{19}$ structure. The atomic
arrangement in $CaAl_{12}O_{19}$ is densely packed and the phase is
nonconducting. Since the sodium linewidth and coupling constant
are unchanged, it is concluded that yttrium does not dissolve
in either the spinel block or the conduction plane to an
appreciable extent.

TABLE 4

Effect of additives in β-alumina on ^{23}Na NMR spectrum at room
temperature. W is the line width and D_Q the quadrupole coupling
constant.

Additive	Concentration (g-atom mole^{-1})	W (gauss)	D_Q (MHz)
None	0	2.0	1.98
Mg	0.30	1.4	2.34
Ni	0.30	1.4	2.30
Li	0.30	1.5	2.32
Ca	0.30	4.0	2.30
Y	0.30	2.0	2.00

   The mobility of sodium is increased by additions of Mg, Ni,
and Li. The enhancement of sodium motion can be explained by the
following tentative model. It is postulated that

   a) Mg^{2+}, Ni^{2+}, and Li^{1+} substitute for Al^{3+} in the spinel
      block.
   b) The electric charge of excess Na^{1+} is compensated by O^{2-}
      interstitials.
   c) The O^{2-} interstitials are located in the conduction plane
      and are held by a strong chemical bond to Al^{3+} inter-
      stitial ions in the spinel block.

The first postulate has been validated by the neutron
diffraction study of β" which established the location of Mg^{2+}
in Al(2) sites in the spinel block. The second is supported by
the change in weight and density of crystals when Na is replaced
by Ag. The third is crystal-chemically attractive because it
completes the oxygen coordination of the interstitial Al^{3+} ion.
The insertion of interstitial O^{2-} in the bridge position between
B-R and anti B-R sites satisfies the oxygen coordination of the
Al^{3+} in the Frenkel defect and also compensates the positive
charge of the excess Na^{1+}. The postulated defect complex is
illustrated in Fig. 7.

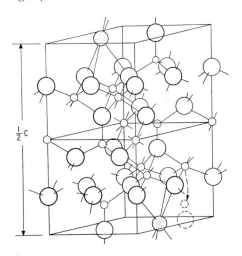

Fig. 7. Frenkel defect in β-alumina. The oxygen coordination of
        the interstitial aluminium (small dashed circle) is com-
        pleted by interstitial oxygen (large dashed circle) in
        the conduction plane.

According to the model, enhancement of the sodium conductivity by monovalent and divalent additives is accomplished by
reduction of the concentration of O^{2-} interstitials in the conduction plane. Approximately one out of ten unit cells in pure β
contains an interstitial O^{2-} ion in the bridge location. Each
intersitial blocks sodium motion between a pair of adjacent sites.
A small concentration will thus have a strong inhibiting effect
on the flow of Na^{1+} ions. The enhancement of the sodium conductivity by Mg^{2+} is a consequence of the smaller number of blocking
O^{2-} ions in the conduction plane. The model also explains the
absence of Frenkel defects, the improved conductivity, and greater

thermal stability of Mg-stabilized β''.

III. TYPE II CONDUCTORS; CALCIA STABILIZED ZIRCONIA

The second category of superionic conductors is the type II or extended defect conductor. Examples are $Ca_xZr_{1-x}O_{2-x}$, $Fe_{1-x}O$, $\alpha-Ag_2HgI_4$, and UO_{2+x}. In contrast to Type III, the activation energy of Type II conductors is not unusual for an ionic solid. The distinguishing characteristic of Type II is that the conductivity is anomalously high for the concentration of charge carriers and the activation energy. With the possible exception of Ag_2HgI_4, Type II conductors contain large concentrations of defects which have coalesced into coherent substructures, or extended defects of submicroscopic dimensions. Structures of this kind are encountered in highly non-stoichiometric compounds and have been termed *Magnéli phases, Guinier-Preston zones* or *shear structures*. The crystallographic details of these highly defective materials are complex and, compared to Type III, relatively little is known about the structural factors which control their conductivity.

The relation of transport properties to structure in Type II conductors has been studied most thoroughly in calcia-stabilized zirconia (CSZ). The name is used to describe a cubic phase in the ZrO_2-CaO systems which extends from about 10 to 20 mole percent CaO. It is an excellent conductor of O^{2-} ions at temperatures above 900°C. X-ray and neutron diffraction, diffusion and conductivity measurements have been made as a function of composition across the solid solution range.

III.1. Idealized Disorder Structure

CSZ crystallizes in the cubic fluorite structure. Ca^{2+} and Zr^{4+} are randomly distributed among the cation sites and electric charge neutrality is maintained by vacancies in the anion sites. The unit cell contains 4 $Ca_xZr_{1-x}O_{2-x}$ molecules with $0.1 < x < 0.2$. The rapid diffusion of O^{2-} ions and the high conductivity is qualitatively explained by the exchange of O^{2-} ions on lattice sites with the large concentration of anion vacancies.

Several aspects of O^{2-} transport in CSZ are not explained by the disordered fluorite model [18].
1) The conductivity is 10 to 100 fold greater than expected according to thermally activated hopping theory. In terms of the free-ion theory, the mean free path and the lifetime in the

free-ion state is of the order of 100 Å and 10^{-12} sec. respectively [4].

2) The composition dependence of the conductivity is anomalous.
The activation energy and the pre-exponential terms in the conductivity equation are strongly dependent of the composition.
3) The resistivity increases with time when CSZ is maintained
at temperatures between 700°C and 1000°C. The high conducting
state is restored by heating to temperatures above 1400°C and
quenching.

III.2. Order-Disorder Transformation

    The anomalous features of ion transport in CSZ are related to
distortions of the fluorite arrangement [18]. The oxygens are
displaced about 0.12 Å from positions they should occupy if the
structure were an ideal fluorite and the conductivity changes
are associated with ordering of oxygen atoms.

    CSZ quenched from high temperature is disordered and diffuse
scattering of neutrons is observed due to short range correlations
between the positions of the oxygen atoms. When CSZ is heated at
700°C to 1000°C for a long time, the diffuse scattering is trans-
formed into sharp Bragg peaks due to long-range ordering of oxygen
ions and vacancies in anion sites and Ca^{2+} and Zr^{4+} in cation
sites. The ordering reaction is sluggish and occurs at a maximum
rate near 1000°C. The crystallography of the atom movements involved in the order-disorder transformation is shown schematically
in Fig. 8.

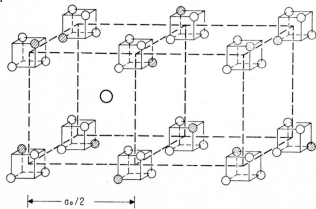

|← —— $a_0/2$ —— →|

Fig. 8.   Oxygen coordination about cation and interstitial sites in
          disordered and ordered CSZ. Large open circle is (Ca,Zr).
          Small open circles are sites statistically occupied by
          oxygen atoms in the disordered phase, the striped circles
          indicate the sites occupied by oxygen in the ordered phase.

III.3. Satellites

The structure in real crystals of CSZ is even more complex
than suggested in the previous section. Satellite reflections are
observed in heavily exposed x-ray and electron diffraction photographs of disordered and ordered CSZ crystals [18]. The
satellites indicate that CSZ probably is not a single phase in
a strict sense and that the defects are organized into sub-
structures which are at least partially periodic. The satellites
are consistent with a periodic variation of electron density
along <112> directions in the crystal with a wavelength of 5.6 Å
and a mean correlation length of 50-100 Å. The relationship between the satellites and the order-disorder transformation discussed in section III.2. is not understood, but it is believed that the
diffraction effects are produced by scattering from substructures
which are responsible for the high conductivity. An ad hoc theory
which relates the ion transport and the satellites is given in
section III.4

III.4. Cooperative Hopping

The extremely long mean free paths calculated for CSZ from
transport data [4] can be understood if the satellites are due
to diffraction from extended defects along which there is
cooperative transport of oxygen. It is instructive to consider
the following simple model. The non-stoichiometric composition
$Ca_xZr_{1-x}O_{2-x}$ can be generated by mixing two fluorite-like cells,
X cells with composition Zr_4O_8 and O cells with composition
$CaZr_3O_7V_o''$ where V_o'' is a doubly ionized oxygen vacancy in the
Kröger-Vink notation [19].

The distribution of runs XXX···, OOO··· and XOX··· was calculated as a function of composition assuming: 1) the cells are
arranged along a linear chain of sites with spacing N_0, 2) the
sites are occupied at random by X and O cells, 3) runs of O
cells can be neglected. The average length of XOX runs as a
function of composition is shown in Fig. 9. The strong composition
dependence of the run length is comparable to that of the pre-
exponential term in the conductivity expression. Alternatively,
in the framework of the free-ion theory, the calculated length
of XOX runs is of the same order as the mean free path which increases from 8 Å at x = 0.1 to 300 Å at x = 0.2.

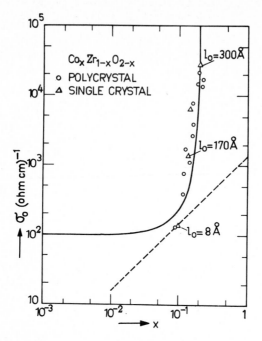

Fig. 9. Length of XOX runs and mean free paths as a function of
composition. The solid line is calculated for the linear
co-hop model, the dashed line is calculated assuming the
pre-exponential to be proportional to x.

IV. SUMMARY

The transport of ions in crystalline materials depends on
the concentration and structure of defects. Considerable progress
has been made in the understanding of ionic conductivity in
solids which contain either very small or very large concentrations of defects. At small concentrations, interactions between
the defects can be neglected. These are termed Type I solids and
the conductivity is given by conventional thermally activated
hopping theory in which the conductivity is proportional to the
concentration, the jump length is of the order of the interatomic
distance, and the activation energy is of the order of the
binding energies of ions in solids. As their concentration increases, the defects interact and coalesce to form Type II or
extended-defect conductors which have complex non-linear properties due to the pseudo two-phase nature of the solid. An
analysis of the structures and transport parameters in the Type II

conductor CSZ shows that the conductivity is highly structure sensitive and it is suggested that there may be cooperative processes involved in the ionic transport. The concentration of defects in Type III conductors is so large that they interact to form quasi-periodic interconnected structures. These are essentially liquid-like conductors which are characterized by the fact that each potentially mobile ion has available at least three vacant neighboring sites of equal, or nearly equal, energy. The microscopic mechanisms for ion transport in the Type III conductor beta-alumina has been investigated by diffraction and NMR methods and many of the important structural and crystal chemical factors which control the conductivity are now understood.

ACKNOWLEDGEMENTS

Part of this work was done in cooperation with staff members of the State University of New York at Albany (SUNYA) and the Brookhaven National Laboratory. In particular, wide-band NMR measurements were made at SUNYA by H.S. Story, D. Kline, and I. Chung[a], and neutron diffraction experiments were made at BNL in cooperation with S.J. LaPlaca, W.C. Hamilton[b], and F. Reidinger[c].

a - Graduate student in Physics, SUNYA.
b - Deceased.
c - Graduate student in Physics, SUNYA.

References

[1] Fast Ion Transport in Solids, Ed. W. van Gool (North-Holland, Amsterdam, 1973).

[2] Extended Defects in Non-Metallic Solids, Eds. L. Eyring and M. O'Keeffe (North-Holland, Amsterdam, 1970).

[3] W.L. Roth, J. Solid State Chem. 4 (1972) 60.

[4] M.J. Rice and W.L. Roth, J. Solid State Chem. 4 (1972) 294.

[5] Y.F. Yao and J.T. Kummer, J.Inorg.Nucl.Chem. 29 (1967) 2453.

[6] C.A. Beevers and M.A. Ross, Z. Krist. 97 (1937) 59.

[7] C.R. Peters, M. Bettman, J.W. Moore and M.D. Glick, Acta Cryst. B27 (1971) 1826.

[8] G. Yamaguchi and K. Suzuki, Bull.Chem.Soc. Japan 41 (1968) 93.

[9] M. Bettman and C.R. Peters, J. Phys. Chem. 73 (1969) 1774.

[10] W.L. Roth, W.C. Hamilton and S.J. LaPlaca, Am. Cryst. Assoc. Abstracts [2] 1 (1973) 169.

[11] J.T. Kummer, Progr. Solid State Chem. 7 (1972) 141.

[12] Y. LeCars, J. Théry and R. Collongues, Rev. Intern. Hautes Températures et Réfractaires 9 (1972) 153.

[13] W.L. Roth and R.J. Romanczuk, J. Electrochem. Soc. 116 (1969) 975.

[14] W.L. Roth and S.P. Mitoff, General Electric Corp. Research and Development Report 71-C-277, (1971).

[15] F. Reidinger, unpublished.

[16] I. Chung, H.S. Story and W.L. Roth, Bull. Philadelphia Meeting of American Physical Society, 25-28 March (1974).

[17] D. Kline, H.S. Story and W.L. Roth, J. Chem. Phys. 57 (1972) 5180.

[18] R.E. Carter and W.L. Roth, in: Electromotive Force Measurements in High-Temperature Systems, Ed. C.B. Alcock (Institute of Mining and Metallurgy, London, 1968).

[19] F.A. Kröger and H.J. Vink, Solid State Phys. 3 (1956) 307.

Crystal Structure and Chemical Bonding in Inorganic Chemistry
Eds. C.J.M. Rooymans and A. Rabenau
© 1975, North-Holland Publishing Company, The Netherlands

THE STABILITY OF THE TRIGONAL-PRISMATIC COORDINATION OF ATOMS IN SOLIDS

C. Haas

Laboratory of Inorganic Chemistry, Materials Science Centre of
the University,

Groningen, The Netherlands

I. INTRODUCTION

One of the aims of solid state chemistry is to understand the relation between the structure, the chemical bonding and the properties of solid compounds. Particularly successful in this respect has been the electrostatic model, initiated by Born [1] and in later years further extended by Van Arkel and De Boer [2,3]. The simple and quantitative considerations of this theory made it possible to understand a large body of structural data in inorganic chemistry. Later, the development of the quantum theory of chemical bonding found extensive application, particularly in organic chemistry [4]. Unfortunately, in practice quantitative quantum mechanical calculations can be carried out only for fairly simple systems. This makes that the old concepts of the ionic model, combined with simple considerations of covalency, are still very useful as a first step in the understanding of the structure and the properties of inorganic solids. Only after a qualitative understanding has been reached in this way, a detailed discussion in terms of more sophisticated quantum mechanical theories is fruitful.

In a simple electrostatic model one expects simple types of coordination, which are usually highly symmetric, such as the octahedral and tetrahedral coordination. However, in many instances the situation is not so simple, and less-symmetric types of coordination are observed. Examples are the asymmetric coordination of anions in layer-type compounds caused by large anion polarizability, linear and square coordination due to covalency, Jahn-Teller distortions, etc. In this paper we will focus our attention on one particular type, the trigonal-prismatic coordination. This type of coordination can not be explained on the basis of the electrostatic model only.

The trigonal-prismatic coordination of metal atoms is found in
MoS_2 and related compounds with a layer-type structure. These
compounds have recently attracted great attention because of
their interesting chemical and physical properties (intercalation
of molecules between the layers, super-conductivity). We will
discuss the structure, chemical bonding and physical properties
of these compounds. Attention will be given particularly to the
question of the stability of the trigonal-prismatic coordination:
why does this type of coordination occur, in which compounds, and
what are the consequences for the physical properties. Finally
we will compare the information about the energy levels of the
electrons, obtained with simple model considerations, with the
results of band structure and self-consistent cluster calculations,
and with direct observations with photoelectron spectroscopy.

II. OCCURRENCE OF THE TRIGONAL-PRISMATIC COORDINATION

In Fig. 1 the trigonal-prismatic coordination of a cation by
six anions is compared with the more common octahedral coordination.

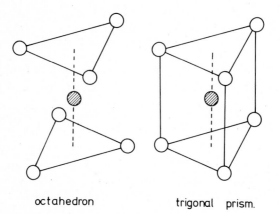

octahedron   trigonal prism.

Fig. 1. Octahedral and trigonal-prismatic coordination of a central
        atom by six ligand atoms.

The trigonal-prismatic coordination can be obtained from the
octahedral one by a rotation of the upper anion triangle over 60°.
In the trigonal prism the anion-anion distances are shorter than
in the octahedron. Therefore, in a simple electrostatic model,
the trigonal-prismatic coordination will have a higher energy. The

basic question which must be answered is why the trigonal-prismatic
coordination occurs nevertheless.

The trigonal-prismatic coordination of transition-metal atoms
occurs in a number of compounds with bidentate sulfur ligands.
An example is given in Fig. 2 [5,6]. Although the considerations
of chemical bonding, to be developped in later sections, also
apply to these complexes, we will not discuss these compounds
further, but restrict our attention to inorganic, nonmolecular
solids.

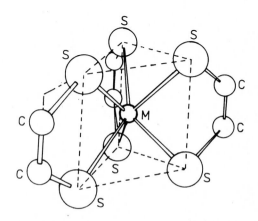

Fig. 2. Trigonal-prismatic complex $M(S_2C_2H_2)_3$, with M = Cr, Mo, W
[5].

The disulfides and diselenides of Nb, Ta, Mo and W, and α-MoTe$_2$,
have structures with a trigonal-prismatic coordination of the
metal atoms [7,8,9]. These compounds have a layer-type sandwich
structure, one sandwich unit consisting of a double layer of
chalcogen atoms with the metal atoms in between. As is well known,
layer structures have a low energy for highly polarizable anions
such as S, Se, Te; the asymmetric coordination of the anion by
the highly charged cations produces a strong polarization of the
anions which stabilizes the structure. The bonding between the
sandwich layers is very weak, and mainly of the Van der Waals type.
This is the reason that these compounds cleave easily. It makes
these materials suitable as solid lubricants: MoS$_2$ is the well-
known non-conducting lubricant "molycote", whereas NbSe$_2$ could
serve as a lubricant which is a good conductor for electricity.

The physical properties are highly anisotropic; the electrical
and thermal conductivities are significantly lower along the c-axis
than they are within the basal plane. Fine powders of MoS_2 and WS_2
have interesting catalytic properties [10].

The disulfides and diselenides of Nb and Ta are metallic and
become superconducting at low temperatures [11]. Because of their
structure, these materials resemble two-dimensional superconductors,
and it was this aspect which stimulated the interest of solid-state
physicists in this group of compounds. The two-dimensional nature
of the superconductivity is demonstrated in a dramatic way by the
properties of the intercalated compounds [12,13,14]. It was found
possible to intercalate a variety of organic molecules in the
Van der Waals gap between the sandwich layers, as shown in Fig. 3.

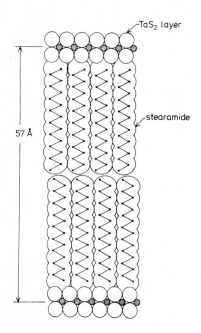

Fig. 3. Intercalation compound of stearamide and TaS_2 [14].

These intercalated compounds are also superconducting with
approximately unchanged transition temperatures. This proves that
the interactions causing superconductivity are within the sandwich
layers.

Some of the dichalcogenides just mentioned also occur in a crystal structure in which the cations have an octahedral coordination of anions. This change of the type of coordination has a large effect on the physical properties. $MoTe_2$, for example, exists in two forms: in the α-form, which is a semiconductor, the Mo atoms have a trigonal-prismatic coordination, whereas in the metallic   β-form the coordination is octahedral. A first-order phase transition between these two forms takes place at 850°C [15]. The resistivity as a function of the reciprocal temperature is sketched in Fig. 4. The physical properties of $1s-TaS_2$ and

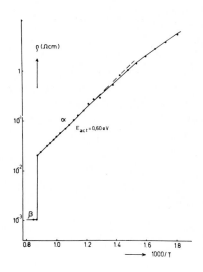

<u>Fig. 4</u>. Metal to semiconductor transition in $MoTe_2$ at 850°C [15].

$1s-TaSe_2$, with an octahedral coordination of the metal atoms, are quite complicated [16,17]. The 1s form of TaS_2 undergoes a metal-semiconductor phase transition at 350°K. Electron diffraction and X-ray studies reveal the presence of a superlattice in $1s-TaS_2$ and $1s-TaSe_2$ [18,19,20].

There exist many other compounds with a trigonal-prismatic coordination of atoms; these will be mentioned only briefly.
(a) It is possible to intercalate alkali metal atoms in the Van der Waals gap of the compounds NbX_2 or TaX_2 (X = S, Se) [21]. In these compounds the environment of the intercalated alkali metals is

either octahedral or trigonal-prismatic. For X = Se trigonal-
prismatic coordination of the alkali metals was found, for X = S
both octahedral and trigonal-prismatic coordination occurs. In
the related compounds A_pTiS_2 (A = Li, Na, K, Cs) the alkali
metals Li and Na have octahedral, K and Cs trigonal-prismatic
coordination [22]. In Na_xVS_2 three different crystal structures
were observed, with the Na atoms occupying octahedral or trigonal-
prismatic sites [23]. The compounds $A_2Pt_4S_6$ (A = K, Rb, Cs) and
isotypes have layer structures with the alkali metals intercalated
between the layers in a trigonal-prismatic coordination [24].
(b) Both metal and nonmetal atoms have a trigonal-prismatic
coordination in NbP, TaP, NbAs and TaAs [8]. The same is true
of phases with the WC-type structure, such as MoP [25], ZrS_{1-x} [7],
$ZrSe_{1-x}$, $ZrTe_{1-x}$ [26], HfS [27], HfSe [28] and possibly TiS_{1-x} [7].
(c) Trigonal-prismatic coordination of anions is also found in
several other structure types, the most common being the NiAs
type and its derivatives, such as the MnP type. Structures of
these types are known for many sulfides, selenides, tellurides
phosphides, arsenides and antimonides of transition metals [8].

These examples show that trigonal-prismatic coordination
is more common for 4d and 5d transition metal atoms than for 3d
transition metal atoms. In nearly all cases the metal atoms with
trigonal-prismatic coordination have a formal configuration d^1 or
d^2. The alkali metal ions K, Rb and Cs and the anions S, Se, Te
can be regarded as d^0 ions. Apparently, the trigonal-prismatic
coordination occurs only for atoms with d^0, d^1 and d^2 configurations.
It is found only in rather covalent compounds, and not in the
highly ionic halides, oxides and nitrides.

III. COVALENCY OF D ELECTRONS

In this section we consider the energy levels of a transition
metal atom in a trigonal-prismatic coordination of anions
[8,29-35]. First we consider the crystal field splitting of the
metal d orbitals. Under the influence of the electrostatic field
of the six negatively charged anions, the d orbitals split into
two doubly-degenerate energy levels e' (m = $\pm$ 2; $d_{x^2-y^2}$, d_{xy}) and
e" (m = $\pm$ 1; d_{xz}, d_{yz}) and a non-degenerate level a_1' (m = 0, d_{z^2})
(m is the magnetic quantum number; the component of the orbital
angular momentum along the trigonal z axis is m$\hbar$ (see Fig. 5).

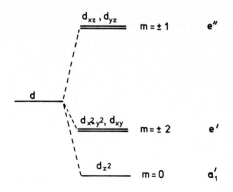

<u>Fig. 5</u>. Crystal field splitting of d orbitals in a trigonal-
prismatic coordination.

It is well known that the simple crystal field model,
although it leads to a qualitative understanding of the splitting
of d levels, does not explain in a quantitative way the properties
of the transition metal compounds. This is especially so for highly
covalent compounds such as MoS_2. For a proper description it is
necessary to take into account covalent mixing of d orbitals
and ligand orbitals. This d-covalency will be considered here
in the most simple manner, by taking into account only σ bonding
between the d orbitals of the central atom and p orbitals of
the ligand anions. The overlap integral V_σ for σ bonding is shown
in Fig. 6. The matrix elements of the d orbitals ϕ_d^m (m = 0, $\pm$ 1,
$\pm$ 2) with linear combinations of the same symmetry ϕ_L^m of ligand
orbitals can be expressed in terms of V_σ ; the results are
given in Table 1. As each d function has a nonvanishing matrix
element with only one combination of ligand functions, the
eigen values for the energy are given by

$$E^m = - \tfrac{1}{2} \Delta \pm \tfrac{1}{2} \left\{ \Delta^2 + 4 \, (V_\sigma^m)^2 \right\}^{\tfrac{1}{2}}$$

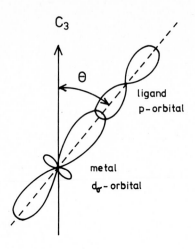

<u>Fig. 6</u>. Overlap integral V_σ for σ bonding.

Table 1

Matrix elements for σ bonding

Symmetry	Matrix element
a_1'	$\langle\phi_d^0 \mid \mathcal{H} \mid \phi_L^0\rangle = \sqrt{3/2}\,(3\cos^2\theta - 1)\,V_\sigma$
a_2''	Non-bonding
e'	$\langle\phi_d^{\pm2} \mid \mathcal{H} \mid \phi_L^{\pm2}\rangle = (3/2)\sin^2\theta\,V_\sigma$
e''	$\langle\phi_d^{\pm1} \mid \mathcal{H} \mid \phi_L^{\pm1}\rangle = 3\sin\theta\cos\theta\,V_\sigma$

The energies of the unperturbed metal d and ligand p orbitals are 0 and $-\Delta$, respectively. The resulting energy levels are given in Table 2. Similar calculations were carried out for a trigonally-distorted octahedron (Table 2).

Table 2

Energy levels for octahedral and trigonal-prismatic coordination, for σ bonding only. Antibonding levels are indicated by an asterisk. $\alpha_1 = 6 \ (V_\sigma/\Delta)^2 (3\cos^2\theta - 1)^2$; $\alpha_2 = 9 \ (V_\sigma/\Delta)^2 \sin^4\theta$; $\alpha_3 = 36 \ (V_\sigma/\Delta)^2 \cos^2\theta \sin^2\theta$.

Trigonal prism	Trigonally-distorted octahedron	Ideal octahedron
$e''^* \ : \ \frac{1}{2}\Delta[-1+(1+\alpha_3)^{\frac{1}{2}}]$	$e_g^* \ : \ \frac{1}{2}\Delta[-1+(1+\alpha_2+\alpha_3)^{\frac{1}{2}}]$	e_g^*
$e'^* \ : \ \frac{1}{2}\Delta[-1 + \{1+\alpha_2\}^{\frac{1}{2}}]$	$a_{1g}^* : \ \frac{1}{2}\Delta[-1 + (1+\alpha_1)^{\frac{1}{2}}]$	
$a_1^* \ : \ \frac{1}{2}\Delta[-1 + (1+\alpha_1)^{\frac{1}{2}}]$	$e_g \ : \ 0$	t_{2g}^*
$a_2'' \ : \ -\Delta$	$a_{2u} : \ -\Delta$	
$a_1' \ : \ \frac{1}{2}\Delta[-1 - (1+\alpha_1)^{\frac{1}{2}}]$	$e_u \ : \ -\Delta$	t_{1u}
$e' \ : \ \frac{1}{2}\Delta[-1 - (1+\alpha_2)^{\frac{1}{2}}]$	$a_{1g} : \ \frac{1}{2}\Delta[-1 - (1+\alpha_1)^{\frac{1}{2}}]$	a_{1g}
$e'' \ : \ \frac{1}{2}\Delta[-1 - (1+\alpha_3)^{\frac{1}{2}}]$	$e_g \ : \ \frac{1}{2}\Delta[-1 - (1+\alpha_2+\alpha_3)^{\frac{1}{2}}]$	e_g

The energy levels are shown in Fig. 7. There are a number of

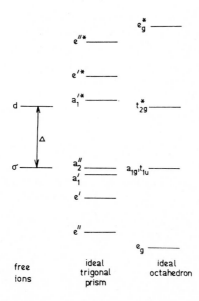

Fig. 7. Energy levels for an ideal trigonal prism $(\cos^2\theta = 3/7)$ and an ideal octahedron $(\cos^2\theta = 1/3)$, for σ bonding only.

bonding and non-bonding levels derived mainly from the ligand
p orbitals; the higher antibonding levels correspond to the metal
d orbitals.

It is possible to estimate the contribution of d-covalency
to the chemical bonding by simple adding the energies of all
occupied energy levels[x]. It is found that the trigonal-prismatic
coordination is more favourable for d-p bonding between cation
and anion than the octahedral coordination (see Fig. 8). For d^0,
d^1 and d^2 (low spin) ions there is a considerable stabilization
of the trigonal-prismatic coordination. For more d electrons the
stabilization gradually disappears. For a d^3 configuration, for
example, all three d electrons are accomodated in the low-lying
t_{2g}^* orbital for the octahedral case. In the trigonal-prismatic
case, however, the configuration will be $(a_1^*)^2 (e'^*)^1$, so that
one electron occupies the higher lying e'^* orbital. The conclusion is that d covalency has a stabilizing effect on the
trigonal-prismatic coordination of atoms with a d^0, d^1 and d^2
(low spin) configuration.

The stabilizing effect of d-covalency depends on V_σ/Δ.
V_σ is a measure of the strength of d-covalency. The energy Δ
is the difference in energy between metal d and ligand p
orbitals, and is directly related to the difference in electronegativity of the two atoms. For strongly ionic compounds
(halides, oxides) Δ is large, and one expects that the
stabilizing effect is small.

[x] It is well known that reliable values for the total energy
cannot be obtained from simple one-electron molecular orbital
calculations, because electron-electron interactions are
not taken into account properly. However, this is not
necessarily so for energy differences. In a comparison of
trigonal-prismatic and octahedral coordination the electron
repulsion energies will be similar for the two systems. The
most important difference will be due to differences in the
anion-anion repulsion. This effect favours the octahedral
coordination.

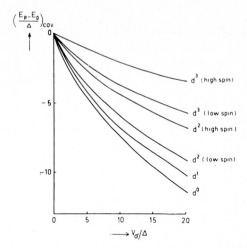

<u>Fig. 8.</u> Contribution of d-covalency to the energy difference
(E_p-E_0) between trigonal-prismatic and octahedral
coördination, as a function of V_σ/Δ and neglecting
electron-electron repulsion.

Because d-covalency is stronger in the trigonal-prismatic
coordination than in the octahedral one, one expects a larger
ionicity in compounds with octahedral coordination. This result
can be obtained from an analysis of the charge distribution
of the occupied molecular orbitals: the positive charge on the
metal atoms is found to be larger for the octahedral coordination.

The considerations given so far are for a single cluster
of atoms MX_6. In a crystal the orbitals of adjacent clusters
interact, which leads to the formation of energy bands. The
bonding and non-bonding levels will form a relatively broad
valence band, the antibonding d levels somewhat narrower d bands.
From these considerations it is easy to understand that the
dichalcogenides of Nb and Ta with a trigonal-prismatic coordination are metallic: for these d^1 ions the lowest d band
(a_1^*) is only half filled. In the dichalcogenides of Mo and W,
with the d^2 ions Mo^{4+} and W^{4+} in a trigonal-prismatic coordination,
this lowest d band is filled, resulting in semi-conducting
properties. For $Mo^{4+}(4d^2)$ in an octahedral coordination (β-$MoTe_2$),
the two d electrons occupy states of the lowest d band derived
from the degenerate t_{2g}^* orbitals; this band can contain six
electrons, so that for β-$MoTe_2$ metallic conductivity is
expected [15]. Thus the simple considerations of d-covalency

lead to a qualitative energy level scheme which explains the
occurrence of metals and semiconductors in the transition metal
dichalcogenides [35].

IV. THE STABILITY OF THE TRIGONAL-PRISMATIC COORDINATION

In this section we discuss the stability of the trigonal-
prismatic coordination with respect to the more common octahedral
coordination.

Electrostatic calculations for MX_2 layer-lattices show that
both the Madelung energy and the polarization energy favour
the octahedral coordination [36]. That the stability of the
trigonal prism is not due to metal-metal bonding is demonstrated
by the complex tris(cis-1,2,-diphenylethene-1,2-dithiolato)rhenium
with a trigonal-prismatic coordination of the rhenium atom [37].
The crystal structure shows that the shortest Re-Re-distances
in the crystal are 9.5 Å, which is much too large for any
appreciable metal-metal bonding.

In the preceding section we have discussed the influence
of d-covalency on the stability of the trigonal-prismatic
coordination. It was found that d-covalency stabilizes the
trigonal-prismatic coordination for d^0, d^1 and d^2 (low spin)
central ions. This is in agreement with the observation that the
trigonal-prismatic coordination is only observed for atoms with
d^0, d^1 and d^2 (low spin) configurations. Because the anions are
further apart in an octahedral coordination, the repulsion
between the anions favours an octahedral coordination.

In Fig. 9 the occurrence of trigonal-prismatic and octahedral
coordination of compounds MX_2 is plotted as a function of the
ratio r^+/r^- of cation and anion radii and f_i [38]. The fractional
ionic character f_i of the M-X bond is defined as
$f_i = 1 - \exp[-\frac{1}{4}(X_m-X_x)^2]$, where X_m and X_x are the electronegativities
of M and X atoms. From Fig. 9 it is found indeed that the
octahedral coordination is obtained for strongly ionic compounds
with small central ions.

The interatomic distances in the transition metal dichalcogenides cannot be interpreted in terms of any unique set of
ionic or covalent radii, the radii will vary as a function
of the electronegativity difference of the bonding atoms.

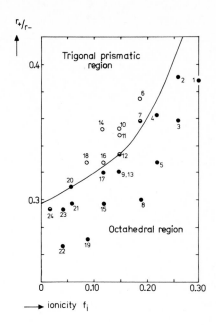

<u>Fig. 9.</u> Occurrence of trigonal-prismatic (open circles) and
octahedral (solid circles) coordination. The half solid
circles represent compound in which both configurations
exist. HfS_2 (1), ZrS_2 (2), $HfSe_2$ (3), TiS_2 (4), $ZrSe_2$ (5),
NbS_2 (6), TaS_2 (7), $TiSe_2$ (8), VS_2 (9), WS_2 (10), $NbSe_2$
(11), $TaSe_2$ (12), $HfTe_2$ (13), MoS_2 (14), VSe_2 (15), WSe_2
(16), $ZrTe_2$ (17), $MoSe_2$ (18), $TiTe_2$ (19), $NbTe_2$ (20),
$TaTe_2$ (21), VTe_2 (22), WTe_2 (23), $MoTe_2$ (24) [38].

In order to account for this effect Gamble [38] defined a new
set of effective radii R_M, R_X with $R_X = \frac{1}{2} \cdot (\text{a-axis})$ and $R_M = R - R_X$.
In terms of these radii a nice separation of the stability
regions of octahedral and trigonal-prismatic coordination is
obtained. This is sketched in Fig. 10. The anions begin to
touch in the direction perpendicular to the layers at $R_M/R_X = 0.527$.
Below the $R_M/R_X = 0.527$ line compounds with a trigonal-prismatic
coordination still occur, but in this region the trigonal
prisms are distorted. For $R_M/R_X < 0.488$ the Born repulsion between
the anions in a trigonal prism would become too large, and in
this region only octahedral coordination is observed.

116   C. Haas

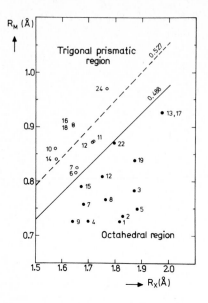

<u>Fig. 10.</u> Trigonal-prismatic and octahedral compounds MX_2 on a R_M vs R_X plot. R_M is the effective cation radius, R_X is the a axis divided by two (numbers as in Fig. 9) [38].

V. PHOTOELECTRON SPECTRA OF MoS_2 AND RELATED COMPOUNDS

Recently it became possible to observe directly the electronic energy levels in molecules and in solids by means of photoelectron spectroscopy. In this type of spectroscopy the material is excited with ultraviolet light or X-rays. The radiation ionizes electrons which enter an electron spectrometer where the kinetic energy E_k of the emitted electrons is measured. The binding energy E_b of the electron in the solid or in the molecule is directly obtained, using the relation $h\nu = E_k + E_b$, where $h\nu$ is the photon energy of the exciting radiation. Photoelectron spectra provide an excellent and very severe test for theories on chemical bonding, and make speculation in this field more difficult.

The photoelectron spectrum of MoS_2 in the region of the valence band is shown in Fig. 11 [39-42]. The band at about 14 eV below the Fermi level E_F is derived mainly from sulfur

3s orbitals, the bands between 2 and 8 eV below E_F are due to
p levels of sulfur, hybridized with metal orbitals. The peak
just below E_F is presumably the narrow band derived from d
orbitals. The spectrum shows that MoS_2 is a semiconductor: there
are no energy levels at E_F.

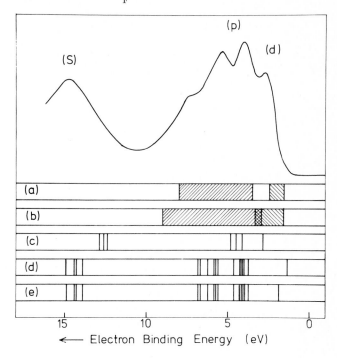

<u>Fig. 11.</u> Photoelectron spectrum of 3R-MoS_2, excited with Al Kα
radiation of hν = 1486.6 eV [39]. The spectrum is compared with calculated energy levels. (a) Mattheiss,
bandstructure [45]; (b) Wood and Pendry, band structure
[48]; (c) Huisman, Mo cluster [35]; (d) De Groot, MST-Xα
[49]; (e) De Groot, MST-Xα + relaxation effects for the
higher levels [49].

The spectra of 1s- and 2s-TaS_2 are given in Fig. 12 [43].
In this compound there is one d electron per Ta atom. In 1s-TaS_2
there are no occupied states just below E_F which corresponds
to the fact that 1s-TaS_2 is a semiconductor. For 2s-$TaS2$, on
the other hand, the Fermi level is situated in the d band,
corresponding to metallic conductivity.

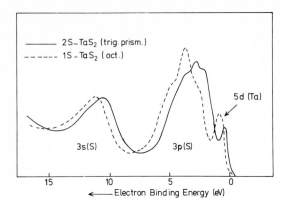

<u>Fig. 12</u>. Photoelectron spectrum of TaS$_2$ [43], excited with
        hν = 1253.6 eV.

With photoelectron spectroscopy it is also possible to
measure the core levels of the atoms. Fig. 13 gives the 4f levels
of Ta in TaS$_2$; the splitting in $4f_{5/2}$ and $4f_{7/2}$ is due to spin-
orbit coupling [43].

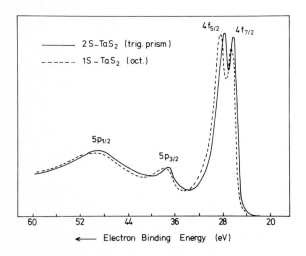

<u>Fig. 13</u>. Photoelectron spectrum of TaS$_2$ [43], excited with
        hν = 1253.6 eV.

In Fig. 14 the infrared reflection spectra of ZrS_2 and MoS_2
are shown [44]. The spectra show a maximum at the so-called
"reststrahlen frequency", due to the vibration of metal atoms
with respect to the nonmetal atoms. The strength of the reflection
maximum is a direct measure of the ionicity of the compound: for
strongly ionic compounds the vibrating atoms have a large charge
and will interact strongly with the infrared radiation. The data
show that the ionicity of ZrS_2 (octahedral coordination) is much
larger than that of MoS_2 (trigonal-prismatic coordination).

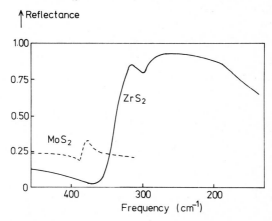

Fig. 14. Infrared reflectance of ZrS_2 and MoS_2, for light polarized
perpendicular to the c-axis. From the spectra dynamic
effective charges $e^* = 4.4$ e for Zr in ZrS_2, and $e^* = 0.6$ e
for Mo in MoS_2 are calculated [44].

This difference in behaviour between ZrS_2 and MoS_2 is due to
the fact that the electronegativity difference between Zr and S
is larger than that between Mo and S (the electronegativities
of Zr, Mo and S are 1.4, 1.8, and 2.5, respectively [4]). Moreover,
the larger d covalency in the trigonal prism will also reduce the
ionicity of MoS_2.

VI. QUANTITATIVE CALCULATIONS OF ELECTRON ENERGY LEVELS

The first quantitative calculations of the energy levels of
MoS_2 were carried out by Huisman [35]. He calculated the energy
levels of a MoS_6 cluster using the semi-empirical Wolfsberg-
Helmholtz molecular orbital method. For the orbital energies the
valence state ionization potentials were used. Numerical Hartree-
Fock functions were used for the atomic orbitals, and the overlap

integrals were calculated directly. Only σ-bonding was taken into
account. In comparing the energy levels of the crystal with
those of the complex, we must leave out the p_π orbitals of
sulfur in the complex. In the crystal each sulfur atom is
surrounded by three Mo atoms. Therefore each p orbital acts as
a σ orbital for one Mo, and nonbonding p orbitals do not
exist in the crystal.

Band structure calculations for MoS_2 and related compounds
have been carried out by several authors [45-48].Mattheiss [45]
employed the augmented plane wave method and used a potential
which is a superposition of atomic potentials derived from
Hartree-Fock calculations for free atoms. The resulting band
structure is shown in Fig. 15. A remarkable result of the calculation is that the lowest d band is hybridized, and has a mixed
$d_z{}^2, d_{x^2-y^2}, d_{xy}$ character.

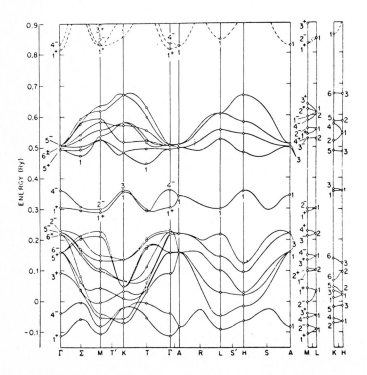

Fig. 15. Band structure of MoS_2, calculated with the Augmented
Plane Wave Method [45].

A drawback of calculations of this type is that they are not self consistent. The band structure calculation is based on a superposition of potentials of neutral atoms. However, in the case of pronounced ionicity, there is a transfer of charge between metal and nonmetal atoms which changes the potential. The relative distance between metal d bands and nonmetal s and p bands is very sensitive to the ionicity and also to the methods employed to take into account exchange and correlation effects. Therefore these band structure calculations based on neutral atom potentials do not give reliable information about the position of the d bands with respect to the valence band. Moreover, a different potential also implies a different radial dependence of the electronic charge on the atoms, and a different radial extension of the metal d orbitals. This will have a large effect on the strength of metal-metal bonding, and on the hybridization of the d band mentioned above.

De Groot [49] has calculated the energy levels of a MoS_6^{8-} cluster using multiple scattering theory. This is a self-consistent ab initio calculation so that no assumptions about the ionicity are necessary [50]. For the calculations the atomic system is divided into muffin-tin spheres. The molecular orbitals are expanded in spherical harmonics in these spheres. From the occupied molecular orbitals the charge density is calculated. This charge density is spherically averaged within the muffin-tin spheres, and   volume averaged in the interstitial space. From these averaged charge densities the electrostatic potential is calculated with Poisson's equation. The contribution of exchange is treated with the $X\alpha$ approximation suggested by Slater, i.e. an exchange potential is assumed proportional to the $1/3$ power of the electronic charge density. With the calculated potential new molecular orbitals are calculated, and a new potential, untill a selfconsistent result is obtained.

Results of the calculations for MoS_2 are compared with the observed photoelectron spectrum, as given in Fig. 11. It is found that all calculations reproduce the main features of the observed spectrum, i.e. a low-lying sulfur 3s band, a valence band (mainly sulfur 3p), and a narrow d band.

For a comparison of photoelectron spectra with energy level calculations, relaxation effects must be taken into account. The final state after the photoemission process has one electron less than the initial state. However, in the final state also

the energy of all other electrons has changed, due to the interaction of these electrons with the hole produced by the photoemission. These relaxation effects are not taken into account in
band theory, but they are readily calculated with the multiple
scattering theory (MST-Xα). From calculations for MoS_2 [49], it
is found that the relaxation effect has a large influence on the
position of the d band with respect to the valence band (Fig. 11).

VII. CONCLUSION

In this paper the structure and bonding of a particular group
of inorganic compounds was discussed. It was attempted to show
that, in order to understand the structure, chemical bonding and
physical properties, the old concepts as ionicity, covalency,
Born repulsion etc., are still very useful.

As a result of the success of the band theory of solids for
the understanding of the electrical transport properties of semiconductors and metals, this theory has become very popular. Indeed,
it is frequently thought that the first thing to know about a solid
is the band structure. However, one should realize that band
theory gives only a one-electron picture of the properties of
electrons, and does not consider electron interactions and correlations properly. The band structure calculations rely heavily
on the long-range periodicity of the crystal. Therefore this
theory is especially suited for describing conduction properties
in metals and semiconductors, because these properties depend
strongly on the periodicity of the crystal.

However, many other properties are of a much more local
character, and depend not in an essential way on the long-range
periodicity. Such properties are the chemical bonding, the type
of coordination, the exchange interactions between magnetic ions,
but also local excitations as are involved in photoelectron
spectra. Indeed in such spectra a photon produces an electron
and a hole close to each other, and the interaction between
electron and hole is a very important effect which should be taken
into account properly. These effects are not treated accurately
in ordinary band theory. Moreover, band theory is not able to
give reliable values for the cohesion energy. Therefore, a discussion of cohesion energy, photoelectron spectra and other local
properties in terms of the band theory is not sufficient, and
the simple electrostatic model as well as calculations for a cluster
of atoms can be helpful.

References

[1] M. Born and A. Landé, Verhandl. Deut. Physik. Ges. 20
(1918) 210;
M. Born, Ann. Physik. 61 (1920) 87.

[2] A.E. van Arkel and J.H. de Boer, Chemische Binding als
Electrostatisch Verschijnsel (Centen, Amsterdam, 1930).

[3] A.E. van Arkel, Moleculen en Kristallen (Van Stockum
's-Gravenhage, 1961).

[4] L. Pauling, The Nature of the Chemical Bond (Cornell Univ.
Press, Ithaca, N.Y., 1941).

[5] G.N. Schrauzer and V.P. Mayweg, J. Am. Chem. Soc. 88
(1966) 3235.

[6] R. Eisenberg, E.I. Stiefel, R. Rosenberg and H.B. Gray,
J. Am. Chem. Soc. 88 (1966) 2874.

[7] F. Jellinek, Arkiv Kemi 20 (1963) 447.

[8] F. Hulliger, in: Structure and Bonding, Berlin 4 (1968) 83.

[9] J.A. Wilson and A.D. Yoffe, Advan. Phys. 18 (1969) 193.

[10] A.L. Farragher and P. Cossee, in: Proc. Fifth Intern.
Congress on Catalysis (North-Holland, Amsterdam, 1972).

[11] M.H. van Maaren and H.B. Harland, Phys. Letters 29A (1969)
571.

[12] F.R. Gamble, F.J. DiSalvo, R.A. Klemm and T.H. Geballe,
Science 168 (1970) 568.

[13] F.R. Gamble, J.H. Osiecki and F.J. DiSalvo, J. Chem. Phys. 55
(1971) 3525.

[14] F.R. Gamble, J.H. Osiecki, M. Cais, R. Pisharody, F.J.
DiSalvo and T.H. Geballe, Science 174 (1971) 493.

[15] M.B. Vellinga, R. de Jonge and C. Haas, J. Solid State Chem.
2 (1970) 299.

[16] A.H. Thompson, F.R. Gamble and J.F. Revelli, Solid State
Commun. 9 (1971) 981.

[17] W. Geertsma, C. Haas, R. Huisman and F. Jellinek, Solid
State Commun. 10 (1972) 75.

[18] J.A. Wilson, F.J. DiSalvo and S. Mahajan, Phys. Rev. Letters
32 (1974) 882.

[19] R. Brouwer and F. Jellinek, Mater. Res. Bull. 9 (1974) 827.

[20] P.M. Williams, G.S. Parry and C.B. Scruby, Phil. Mag. 29
(1974) 695.

[21] W.P.F.A.M. Omloo and F. Jellinek, J. Less-Common Metals 20
(1970)121.

[22] A. Leblanc-Soreau, M. Danot, L. Trichet and J. Rouxel,
      Mater. Res. Bull. 9 (1974) 191.

[23] G.A. Wiegers, R. van der Meer, H.H. van Heiningen, H.J.
      Kloosterboer and A.J.A. Alberink, Mater. Res. Bull., to be
      published.

[24] M. Rüdorff, A. Stössel and V. Schmidt, Z. Anorg. Allg. Chem.
      357 (1968) 264.

[25] S. Rundqvist and T. Lundström, Acta Chem. Scand. 17 (1963)
      37.

[26] H. Hahn and P. Ness, Z. Anorg. Allg. Chem. 302 (1957) 37,136.

[27] H.F. Franzen and J. Graham, J. Inorg. Nucl. Chem. 28 (1966)
      377.

[28] H.F. Franzen, private communication.

[29] C. Creveceur, Thesis, Leiden (1964).

[30] K. Koerts, Thesis, Leiden (1965).

[31] J.B. Goodenough, Mater. Res. Bull. 3 (1968) 409.

[32] R. Huisman, Thesis, Groningen (1969).

[33] R. de Jonge, Thesis, Groningen (1970).

[34] K. Anzenhofer, J.M. van den Berg, P. Cossee and J.N. Helle,
      J. Phys. Chem. Solids 31 (1970) 1057.

[35] R. Huisman, R. de Jonge, C. Haas and F. Jellinek, J. Solid
      State Chem. 3 (1971) 56.

[36] R.J. de Munk, Afstudeerverslag, Technical University, Delft
      (1957).

[37] R. Eisenberg and J.A. Ibers, Inorg. Chem. 5 (1966) 411.

[38] F.R. Gamble, J. Solid State Chem. 9 (1974) 358.

[39] G.K.Wertheim, F.J. DiSalvo and D.N.E. Buchanan, Solid State
      Commun. 13 (1973) 1225.

[40] J.C. McMenamin and W.E. Spicer, Phys. Rev. Letters 29 (1972)
      1501.

[41] P.M. Williams and R.F. Shepherd, J. Phys. C6 (1973) L36.

[42] R.H. Williams, J.H. Thomas, M. Barber and N. Alford, Chem.
      Phys. Letters 17 (1972) 142.

[43] R. Eppinga, private communication.

[44] G. Lucovsky, R.M. White, J.A. Benda and J.F. Revelli, Phys.
      Rev. B7 (1973) 3859.

[45] L.F.Mattheiss, Phys. Rev. B8 (1973) 3719.

[46] R.V. Kasowski, Phys. Rev. Letters 30 (1973) 1175.

[47] R.A. Bromley, R.B. Murray and A.D. Yoffe, J. Phys. C5 (1972)
      738, 746, 759, 3038.

[48] K. Wood and J.B. Pendry, Phys. Rev. Letters 23 (1973) 1400.

[49] R.A. de Groot, private communication.

[50] K.H. Johnson, J. Chem. Phys. 45 (1966) 3085.

Crystal Structure and Chemical Bonding in Inorganic Chemistry
Eds. C.J.M. Rooymans and A. Rabenau
© 1975, North-Holland Publishing Company, The Netherlands

THE MADELUNG PART OF THE LATTICE ENERGY, MAPLE, AS A GUIDE IN
CRYSTAL CHEMISTRY

Rudolf Hoppe
University of Giessen,
Institut für Anorganische und Analytische Chemie,
63, Giessen, Germany

I. INTRODUCTION

In Solid State Chemistry one of the fundamentals in understanding chemical and physical properties is the crystal structure.
This term is not well defined. Here crystal structure means the
result of a three-dimensional periodic repetition of the unit cell
only. The lattice space is infinite. Imperfections like disorder
or dislocations are neglected. Even then a "real understanding"
of *Long Range Order* (LRO) is difficult:

Despite the ability of Man to deal with infinity in the
mathematical way using his intelligence, Man has indeed serious
difficulties to grasp infinity mentally - perhaps due to his
biological background of evolution.

II. ON THE DESCRIPTION OF LONG RANGE ORDER (LRO)

There are various possibilities to characterize the
individual position of the crystallographically different
particles within the infinite lattice space of a given structure,
for instance:

A) the chosen particle can be connected with all other ones
(of the same sort only or of all sorts) thus generating an
infinite set, a "bushel" of distance vectors. Unfortunately
this is as difficult to oversee mentally as the infinite
lattice itself.

B) one can replace each of these vectors by one of the same
direction but of reciprocal length. The result is a *"hedgehog"*
(fig. 1) of *finite size* and *unchanged symmetry*. Unfortunately
this interesting inversion is not studied systematically yet.

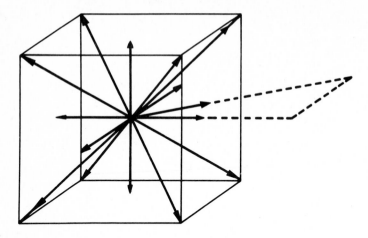

Fig. 1. The hedgehog of reciprocal vectors in case of CsCl.

The problem is much simpler if the vector character of the
reciprocal distances is omitted and the set of lengths only
regarded. Then we can use Madelung Constants (MF) or the
corresponding *Madelung Part* of the *Lattice Energy* E_M (MAPLE),
which due to the law of Coulomb contains the set of distances
in the form of $1/d$ values, thus dealing with the Lattice
Energy, E_L:

$$E_L = E_M + E_B + F_D + E_P + E_C \cdots + E_{corr},$$

where E_{corr} is needed to safe equality of both sides of the
equation, E_C corrects for covalency, E_P for polarizibility,
E_D for forces of dispersion, E_B for Born's repulsion. E_M, MAPLE,
would be the whole Lattice Energy if all ions were charged
points only.

III. SOMETHING ABOUT MAPLE

Therefore we tried to use MAPLE (kcal/mol) as a new guide
to crystal chemistry in the hope LRO could be caught within a
nutshell [1,2,3].

It should be remembered that a Madelung Constant MF depends
on the charges which have been assumed (n_j) (often the numbers
of oxidation states adopted) and of the distance of reference
d_s (here, if not otherwise noted, the shortest of all distances).
MF is always the sum of *Partially Madelung Constants* , PMF:

$$MF(Na^+Cl^-) = PFM(Na^+) \qquad + PMF(Cl^-)$$
$$1.74756... = 1x\ 0.87378. + 1\ x\ 0.87378...$$

or

$$MF(Ca^{2+}F_2^-) = PMF(Ca^{2+}) \qquad + 2\ x\ PMF(F^-)$$
$$5.03878... = 3.27612... \qquad + 2\ x\ 0.88133..$$

Correspondingly we have

$$MAPLE(Na^+Cl^-) = \qquad MAPLE(Na^+) + MAPLE(Cl^-)$$
$$205.6 \qquad = \qquad 102.8 \qquad + \quad 102.8 \quad (kcal/mol)$$
$$MAPLE(Ca^{2+}F_2^-) = \qquad MAPLE(Ca^{2+}) + 2\ x\ MAPLE(F^-)$$
$$706.8 \qquad = \qquad 459.5 \qquad + 2\ x\ 123.6 \quad (kcal/mol)$$

because e.g.:

$MAPLE(Na^+Cl^-) = MF(Na^+Cl^-)\ x\ 331.81/2.819(kcal/mol)$ as well as

$MAPLE(Na^+) \quad = PMF(Na^+) \quad x\ 331.81/2.819(kcal/mol)$,

in which 331.81 is a conversion constant and 2.819 the shortest Na^+-CL^- distance in Å units.

MAPLE of ions like Ca^{2+} can be reduced to unit charge and unit distance thus enabling us to compare structures of different spacing and different assumed charges,

$$_1^*MAPLE \qquad = \frac{MAPLE}{|^n j|^2}\ x\ d_s,$$

e.g. $_1^*MAPLE(Ca^{2+}) \qquad = \frac{459.5}{4}\ x\ 2.365_5 = 271.8$

If this reduction is done not with respect to the shortest distance d_s but to another one (d_i) we have

$$_1^*MAPLE \quad = \frac{MAPLE}{|^n j|^2}\ x\ d_i$$

The range of MF of A^+B^- types ranges from 1.76 (typical ionic) to 1.00 (typical covalent), see Table 1.

TABLE 1

MF-values of AB-types

Type	MF	Type	MF
CsCl	1.76267...	PbO red	1.50699...
NaCl	1.74756...	PbO yellow	1.44879...
CuCl	1.63805...	HgS	1.49503...
ZnS	1.64132...	α-HgO	1.47273...
TlF	1.64014...	β-HgO	1.4802....
TlI	1.63029...	ICl	1.1102....
		AuI	1.0208....

IV. MAPLE - A SIMPLE INDICATOR OF LRO?

We have calculated MAPLE starting with Na_g^+ and Cl_g^- (fig. 2), stepwise, in forming first a line [100] , than a sheet (100) and at last the 3-dimensional lattice of NaCl. The values indicate that MAPLE unfortunately is not very sensitive to LRO. This is drastically confirmed by the values of fig. 3 which indicate that even a triple layer of NaCl-type already has "inside" MAPLE values identical with that of the NaCl-type itself.

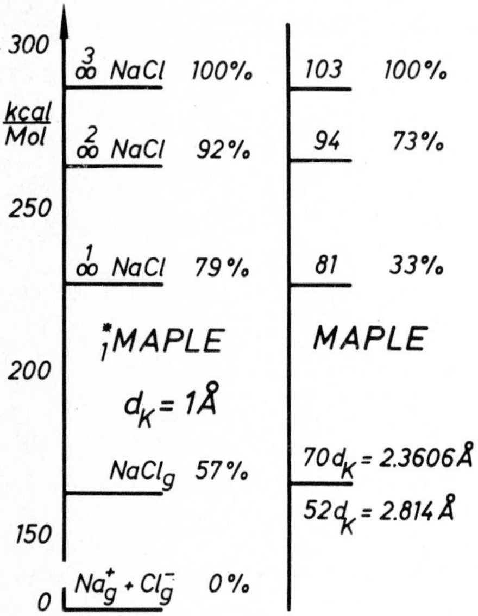

Fig. 2. MAPLE of Na^+

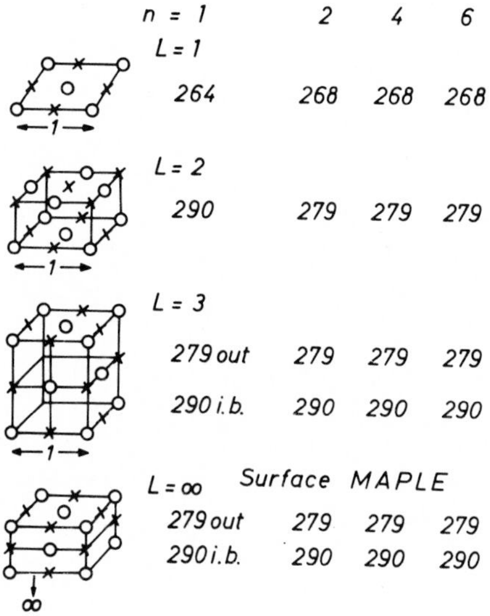

Fig. 3. $^*_1$MAPLE of NaCl (kcal/mol)
n: c'/c; c': elongated
c-axis

Other examples, mentioned below, confirm this in general. The reason is obviously that the NaCl lattice can be regarded as composed of an infinite series of e.g. such triple layers, each of them playing the role of a *multipole*. This holds for other structure types too.

It is now understood why Q_i [3] (in case of simple structures the sum of the assumed charges of the next neighbours divided by the charge of the particle under inspection) governs the reduced MAPLE values: $^*_1$MAPLE, see [3] and Table 2. Thus in general MAPLE is no good indicator of LRO.

MAPLE in Crystal Chemistry

TABLE 2

$$\overset{*}{_1} MAPLE = f\ (Q_i)$$

Type	C.N.	Q_i	$\overset{*}{_1}$MAPLE (kcal/mol)
Cs^+Cl^-			
Cs^+:	8	8	292
Na^+Cl^-			
Na^+:	6	6	290
Cu^+Cl^-	4	4	272
–	3	3	255
–	2	2	230
Example:			
CaF_2			
Ca^{2+}:	8	4	272 !
F^-	4	8	292 !

V. ON THE PRACTICAL USE OF MAPLE

Nevertheless MAPLE is still a useful tool in dealing with and even a guide to crystal chemistry [4,5].

A) MAPLE of different modifications

Table 3.1-31 shows a compilation of MAPLE values of modifications of binary compounds, Table 4.1-3 gives a small selection out of a lot of dimorphic, ternary compounds. In all cases where the structures are well elucidated we find a good agreement, the difference in MAPLE values then $\leqslant 1\%$. In this sense Table 3.19 and 3.20 is e.g. a good guide to the characteristic structural problems of SiO_2 yet remaining unsolved[*].

---

[*] In the Tables the abreviations have the following meaning:
HT: high temperature modification
LT: low temperature modification
HP: high pressure modification
LP: low pressure modification.

<u>TABLE 3</u>

<u>MAPLE</u> (kcal/mol) of polymorphic compounds

Table 3.1

	TlF (orthorh.) [6]	TlF(tetrag.) [7]
Tl(1)$^+$	104.9	104.1
Tl(2)$^+$		104.1
F(1)$^-$	104.9	104.1
F(2)$^-$		104.1
Σ	209.8	208.2
Δ		-0.8%

Table 3.2

	α-PbF$_2$ [8]	β-PbF$_2$ [9]
Pb^{2+}	421.8	423.5
F(1)$^-$	115.8	113.9
F(2)$^-$	104.4	
Σ	642.0	651.3
Δ		+1.4%

Table 3.3

	α-UF$_5$ [10]	β-UF$_5$ [11]
U^{5+}	2333.1	2385.1
F(1)$^-$	112.8	137.3
F(2)$^-$	178.6	157.9
F(3)$^-$		85.0
Σ	2962.9	3008.3
Δ		+1.5%

Table 3.4 [12]

	$\alpha-ZnCl_2$	$\beta-ZnCl_2$	$\gamma-ZnCl_2$
$Zn(1)^{2+}$	378.7	384.6	378.3
$Zn(2)^{2+}$		373.9	
$Zn(3)^{2+}$		373.6	
$Cl(1)^-$	117.9	113.5	119.2
$Cl(2)^-$		121.0	
$Cl(3)^-$		120.8	
$Cl(4)^-$		114.7	
$Cl(5)^-$		122.2	
$Cl(6)^-$		114.9	
Σ	614.5	613.1	616.7
Δ		- 0.2%	+ 0.4%

Table 3.5 [13]

	$CfCl_3$(orthorh.)	$CfCl_3$(hex)
Cf^{3+}	744.1	754.3
$Cl(1)^-$	95.3	
$Cl(2)^-$	109.5	107.1
Σ	1058.5	1075.6
Δ		+ 1.6%

Table 3.6 [14]

	$CrCl_3$(monocl.)	$CrCl_3$(hex.)
Cr^{3+}	787.2	900.3
$Cl(1)^-$	129.5	116.3
$Cl(2)^-$	132.5	
Σ	1181.7	1249.2
Δ		+ 5.7%

Table 3.7

	$\alpha-WCl_6$ [15]	$\beta-WCl_6$ [16]
$W(1)^{6+}$	3008.0	2994.9
$W(2)^{6+}$		2954.2
$Cl(1)^-$	143.9	142.3
$Cl(2)^-$		131.3
$Cl(3)^-$		148.3
Σ	3871.4	3811.6
Δ		- 1.5%

Table 3.8

	EuI_2(monocl.)[17]	EuI_2(orthorh.) [18]
Eu^{2+}	311.3	314.7
$I(1)^-$	82.7	82.7
$I(2)^-$	91.1	91.0
Σ	485.1	488.4
Δ		+ 0.7%

Table 3.9   [19]

	HgI_2, red	HgI_2, yellow
Hg^{2+}	308.9	288.2
$I(1)^-$	93.5	92.3
$I(2)^-$		92.6
Σ	495.9	473.1
Δ		- 4.6%

Table 3.10

	α-PdI_2 [20]	β-PdI_2 [21]
Pd^{2+}	349.1	342.4
$I(1)^-$	107.2	107.8
$I(2)^-$		104.0
Σ	563.5	554.2
Δ		- 1.7%

Table 3.11

	BeO(LT) [22]	BeO(HT) [23]
Be^{2+}	660.8	639.5
O^{2-}	660.8	648.0
Σ	1321.6	1287.5
Δ		- 2.6%

Table 3.12

	$\alpha\text{-Bi}_2\text{O}_3$ [24]	$\beta\text{-Bi}_2\text{O}_3$ [25]
$Bi(1)^{3+}$	1027.2	1051.3
$Bi(2)^{3+}$	1030.5	
$O(1)^{2-}$	478.1	464.6
$O(2)^{2-}$	471.9	446.7
$O(3)^{2-}$	461.5	
Σ	3469.2	3478.5
Δ		+ 0.3%

Table 3.13

	$\text{GeO}_2(\text{quarz})$ [26]	$\text{GeO}_2(\text{T-cristob.})$ [27]	$\text{GeO}_2(\text{rutile})$ [28]
Ge^{4+}	2063.0	2016.5	2147.5
O^{2-}	647.7	646.3	620.1
Σ	3358.4	3309.1	3387.7
Δ		- 1.5%	+ 0.9%

Table 3.14

	$\alpha\text{-HgO}$ [29]	$\beta\text{-HgO}$ [30]
Hg^{2+}	483.8	483.2
O^{2-}	483.6	483.2
Σ	967.4	966.4
Δ		- 0.1%

Table 3.15

	PbO, red [31]	PbO, yellow [32]
Pb^{2+}	475.1	473.2
O^{2-}	392.1	394.0
Σ	867.2	867.2
Δ		$\pm$ 0%

Table 3.16

	$\text{H-Nb}_2\text{O}_5$ [33]	$\text{N-Nb}_2\text{O}_5$ [34]	$\text{M-Nb}_2\text{O}_5$ [35]	$\text{B-Nb}_2\text{O}_5$ [36]	$\text{R-Nb}_2\text{O}_5$ [37]
Σ	8985.7	8924.5	8888.2	8993.8	9042.1
$\Delta(\%)$		- 0.7	- 1.1	+ 0.1	+ 0.6

Table 3.17

	SnO(orthorh.)[38]	SnO(tetrag.)[39]
$Sn(1)^{2+}$	476.7	495.5
$Sn(2)^{2+}$	484.7	
O^{2-}	304.4	408.5
Σ	785.1	904.0
Δ		+15.1%

Table 3.18

	α-PtO$_2$[40]	β-PtO$_2$ [41]
Pt^{4+}	1878.1	2048.4
O^{2-}	403.1	585.3
Σ	2684.3	3219.0
Δ	-16.6%	

Table 3.19

	α-quarz[42]	β-quarz [43]	cristobalite (LT)[45]	cristobalite (HT)[44]
$Si(1)^{4+}$	2229.4	2208.9	2220.9	2212.0
$Si(2)^{4+}$				2191.2
$O(1)^{2-}$	709.7	703.4	720.8	786.2
$O(2)^{2-}$				687.6
Σ	3648.8	3615.7	3662.5	3626.1
Δ (%)		- 0.9	+ 0.4	- 0.6

Table 3.20

	tridymite (LT) [46]	tridymite (HT) [47]	keatite [48]	stishovite [49]	coesite [50]
$Si(1)^{4+}$	2309.0	2358.2	2263.8	2273.4	2263.4
$Si(2)^{4+}$			2219.2		2216.6
$O(1)^{2-}$	737.6	913.6	724.6	658.9	668.4
$O(2)^{2-}$	736.4	729.9	694.7		687.9
$O(3)^{2-}$	752.5		737.4		709.3
$O(4)^{2-}$					700.9
$O(5)^{2-}$					707.9
Σ	3798.5	3909.9	3686.7	3591.2	3638.1
Δ(%)	+ 4.1	+ 7.2	+ 1.0	- 1.6	- 0.3

Table 3.21

	TeO_2(orthorh.) [51]	TeO_2(tetrag.) [52]
Te^{4+}	1891.1	1898.5
$O(1)^{2-}$	511.2	532.2
$O(2)^{2-}$	552.5	
Σ	2954.8	2962.9
$\Delta(\%)$		+ 0.3

Table 3.22

	rutile [53]	brookite [54]	anatase [55]	TiO_2(HP) [56]
Ti^{4+}	2060.9	2052.4	2044.9	2079.3
$O(1)^{2-}$	596.2	597.3	610.1	592.3
$O(2)^{2-}$		603.9		
Σ	3253.2	3253.6	3265.1	3263.9
$\Delta(\%)$		+ 0.0	+ 0.4	+ 0.3

Table 3.23

	Rh_2O_3(LT,LP) [57]	Rh_2O_3(HT,HP) [58]	Rh_2O_3(HT,LP) [59]
$Rh(1)^{3+}$	1178.5	1187.6	1239.9
$Rh(2)^{3+}$			1154.5
$O(1)^{2-}$	568.6	559.0	610.4
$O(2)^{2-}$		561.8	578.4
$O(3)^{2-}$			512.4
Σ	4062.8	4055.0	4095.6
$\Delta(\%)$		- 0.2	+ 0.8

Table 3.24 [60]

	$SO_3(\beta)$	$SO_3(\gamma)$
$S(1)^{6+}$	4512.7	4149.3
$S(2)^{6+}$		4621.1
$S(3)^{6+}$		4898.9
$O(1)^{2-}$	953.8	288.4
$O(2)^{2-}$	743.0	837.7
$O(3)^{2-}$	751.7	617.1
$O(4)^{2-}$		976.9
$O(5)^{2-}$		990.1
$O(6)^{2-}$		917.7
$O(7)^{2-}$		838.7
$O(8)^{2-}$		961.0
$O(9)^{2-}$		590.9
Σ	6961.2	6895.9
Δ		- 0.9%

Table 3.25

	Sb_2O_3(senarmontite)[61]	Sb_2O_3(valentinite) [62]
Sb^{3+}	1099.0	1091.3
$O(1)^{2-}$	453.5	502.9
$O(2)^{2-}$		451.6
Σ	3558.5	3588.7
Δ		+ 0.8%

Table 3.26

	WO_3(LT)[63]	WO_3(RT) [64]	WO_3(HT)[65]
$W(1)^{6+}$	4391.9	4292.4	4306.6
$W(2)^{6+}$		4316.5	
$O(1)^{2-}$	647.0	637.8	603.9
$O(2)^{2-}$	682.8	656.7	658.9
$O(3)^{2-}$	653.4	635.9	
$O(4)^{2-}$		648.0	
$O(5)^{2-}$		630.2	
$O(6)^{2-}$		646.6	
Σ	6375.1	6232.1	6228.0
$\Delta(\%)$	+ 2.3		- 0.1

Table 3.27

	R-P_2O_5 [66]	S-P_2O_5 [67]	M-P_2O_5 [68]
P(1)$^{5+}$	3257.8	3345.0	3118.7
P(2)$^{5+}$		3322.7	3109.4
O(1)$^{2-}$	771.2	843.2	665.8
O(2)$^{2-}$	787.2	631.7	844.1
O(3)$^{2-}$	703.9	680.2	869.3
O(4)$^{2-}$		800.9	667.7
Σ	10269.0	10424.6	10128.0
Δ(%)		+ 1.5	- 1.4

Table 3.28

	Cu_2S(LT)[69]	Cu_2S(HT) [70]
Σ(Δ)	677.3	676.1 (= - 0.2%)

Table 3.29

	α-$ThBr_4$ [71]	β-$ThBr_4$ [72]
Th^{4+}	1189.6	1192.1
Br^{-}	99.0	97.9
Σ	1585.6	1583.7
Δ(%)		- 0.1

Table 3.30 [73]

	As_2O_3(arsenolithe)	As_2O_3(claudetite)
As(1)$^{3+}$	1201.0	1200.1
As(2)$^{3+}$		1082.8
O(1)$^{2-}$	518.0	458.1
O(2)$^{2-}$		504.9
O(3)$^{2-}$		531.6
Σ	3956.0	3777.5
Δ(%)		- 4.5

Table 3.31

	$B_2O_3(I)$ [74]	$B_2O_3(II)$ [75]
$B(1)^{3+}$	1475.7	1473.8
$B(2)^{3+}$	1467.9	
$O(1)^{2-}$	765.3	728.4
$O(2)^{2-}$	754.6	782.5
$O(3)^{2-}$	774.0	
Σ	5237.5	5241.0
Δ (%)		+ 0.1

TABLE 4

MAPLE (kcal/mol) of a few ternary compounds.

Table 4.1

$BaCu_4S_3$	binary [76,77]	ternary	
		α-form [78]	β-form [78]
Ba^{2+}	363.1	342.7	355.6
$S(1)^{2-}$	363.1	402.7	402.2
$S(2)^{2-}$	443.6*	420.6	
$S(3)^{2-}$	443.6*	422.6	430.1
$Cu(1)^+$	116.9*	131.0	126.4
$Cu(2)^+$	116.9*	120.5	
$Cu(3)^+$	116.9*	119.5	106.3
$Cu(4)^+$	116.9*	133.6	
$\Sigma(\Delta)$	2081.0	2093.2(= +0.6%)	2055.5(= -1.2%)

* average

Table 4.2

γ-LiTbO$_2$	binary [79,80]	ternary [81]	Δ
Tb^{3+}	1057.7*	1019.4	+ 38.4
$O(1)^{2-}$	543.5	527.5	+ 16.0
$O(2)^{2-}$	488.7*	491.1	- 2.4
Li^+	146.2	178.4	- 32.2
Σ	2236.1	2216.4	+ 19.8
Δ			(= 0.9%)

* average (Tb_2O_3)

Table 4.3

| RbFeF$_4$ | binary [82,83] | ternary | |
		α-form [84]	β-form [85]
Rb$^+$	102.8	68.8	113.8
Fe^{3+}	1067.0	998.9	1042.1
F(1)$^-$	102.8	184.9	170.6
F(2)$^-$	158.6	198.5	170.6
F(3)$^-$	158.6	148.3	118.3
F(4)$^-$	158.6	148.3	118.3
Σ	1748.4	1747.7	1733.7
Δ(%)		0.0	- 0.8

B) MAPLE of ternary compounds

We have calculated MAPLE for a lot of ternary fluorides
and oxides as well as chlorides, bromides, iodides and
sulfides. Similar to what has been found in the case of
different modifications, MAPLE judges the comparibility of
the structure of the binary compounds with that of the
ternary one. In general the difference is as in the cases
mentioned above ≤ 1%. For some examples see Table 5 and
Table 6. In case of BaZnO$_2$, Table 7 shows a comparison
between "expected" and "observed" values of $^*_1$MAPLE [3] .

The case of "old" and "new" Ba$_2$TiO$_4$ in which we have
been interested with respect to Rb$_2$TiO$_3$ with a C.N. 4 of
Ti^{4+} (see below) is shown in Table 8. The MAPLE values of
Ti^{4+} indicated very clearly that at least the shortest
Ti-O distance (old value: 1.64 Å) is too small (new value:
1.76$_6$ Å).

TABLE 5

MAPLE (kcal/mol) of ternary compounds

binary	ΣMAPLE		ternary ΣMAPLE	Δ MAPLE
BaS	726.2 ⎫			
MnS	888.0 ⎭ 1614.2		$BaMnS_2$ [86] 1626.9	+12.7 = +0.8%
BaS	726.2 ⎫			
Cu_2S(LT) [77]	677.2 ⎭ 1403.3		$BaCu_2S_2$ [87] 1412.0	+ 8.6 = +0.6%
BaF_2	622.8 ⎫			
CuF_2 [88]	790.9 ⎭ 2036.5		Ba_2CuF_6 [89] 2023.0	-13.5 = -0.7%
CaF_2	706.8 ⎫			
CrF_3 [90]	1555.6 ⎭ 2262.4		$CaCrF_5$ [91] 2261.8	- 0.6 = -0.0%
LiF	288.7 ⎫			
SbF_5 [92,93]	3437.0 ⎭ 3725.7		$LiSbF_6$ [94] 3769.6	+43.9 = +1.2%

TABLE 6

$BaTiO_3$: MAPLE (kcal/mol)

	binary	orthorhombic [95]	hexagonal [96]
Ba^{2+}	420.	446.	423.
			445. (2x)
Ti^{4+}	2062.	2055.	2162.
			1981. (2x)
O^{2-}	420.	536.	561.
	597. (2x)	537. (2x)	530. (2x)
Σ	4096.	4111.	4111.

TABLE 7

$BaZnO_2$: MAPLE (kcal/mol) [97]

	MAPLE	Q	$_i^*MAPLE_{obs.}$	expected by Q_i
Ba^{2+}	407.	5.2	267.	280.
Zn^{2+}	580.	3.8	285.	270.
O^{2-}	480.	2.2	236.	235.
Σ	1947.			
ΣBaO+ZnO	1947.			

TABLE 8

MAPLE of Ba_2TiO_4 (kcal/mol)

	"old" [98]	"new" [99]
Ba^{2+} (1)	423.6	403.1
Ba^{2+} (2)	408.1	451.6
Ti^{4+}	2159.0	2001.9
O^{2-} (1)	510.4	530.9
O^{2-} (2)	526.7	518.0
O^{2-} (3)	530.1	516.8
O^{2-} (4)	556.6	517.5
Σ	5114.5	4939.8
Σ binary*	4934.1	4934.1 *TiO_2 = rutile, with MAPLE(Ti^{4+}) = 2062.
Δ	+108.4	+ 5.7
$\Delta(\%)$	+ 3.6$_6$	+ 0.1$_2$

VI. MAPLE IN CASE OF NEW TERNARY OXIDES OF METALS "RICH IN CATIONS"

During the last   years we have prepared some new oxides
"rich in cations" with unusual crystal structures. For example
KAgO [100] with square groups (Ag_4O_4), K_2NiO_2 [101] with XeF_2-
like (NiO_2)-groups, Li_2NiO_2 [102] with $(NiO_{4/2})$-chains, Li_8CoO_6
[103] and Li_4CoO_4 [104] with (CoO_4)-tetrahedrons, $K_6(Co_2O_7)$ [105]
and $Na_8(Ga_2O_7)$ [106] with sorosilicate-like groups, Rb_2CoO_3 [107]
and Rb_2TiO_3 [108] with C.N. 4 and Zweier-Einer-chains,
$K_6(Fe_2O_6)$ [109] with Al_2Cl_6-like groups, $Na_4(FeO_3)$ [110] the
first oxoferrate (II) with nearly planar CO_3-like groups,
$Na_6(ZnO_4)$ [111] and $Na_5(GaO_4)$ [112] with tetrahedrons and
$K_4(Be_2O_4)$ [113] with planar groups showing C.N. 3 for Be. In
all these and other cases of unusual coordination we have
calculated MAPLE to test the validity of MAPLE. The agreement
is very good even here, see the figures 4-9 and Tables 9-13.

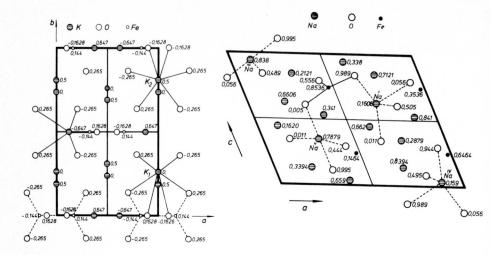

Fig. 4. Structure of $K_6(Fe_2O_6)$　　**Fig. 5.** Structure of Na_4FeO_3

TABLE 9

MAPLE (kcal/mol) of $K_6(Fe_2O_6)$ [109]

$K_6(Fe_2O_6)$		binary	ternary	Δ
K^+	(3x)	105.0	111.6[*]	+ 19.8
Fe^{3+}	(1x)	1199.7	1126.8	- 72.9
$O(1)^{2-}$	(1x)	390.1[**]	551.9	+161.8
$O(1)^{2-}$	(1/2x)	390.1[**]	476.8	+ 43.4
$O(2)^{2-}$	(3/2x)	573.0[***]	476.8	-144.3
Σ		2959.3	2967.1	+ 7.8 = +0.25%

[*] average of K(1), K(2), K(3): 104.8 - 121.6 - 109.3
[**] in K_2O;　[***] in α-Fe_2O_3, the sequence is arbitrary

TABLE 10
MAPLE (kcal/mol) of $Na_4(FeO_3)$ [110]

$Na_4(FeO_3)$		binary	ternary	Δ
$Na(1)^+$	(1x)	121.7	132.1	+ 10.4
$Na(2)^+$	(1x)	121.7	122.2	+ 0.5
$Na(3)^+$	(1x)	121.7	129.5	+ 7.8
$Na(4)^+$	(1x)	121.7	130.7	+ 9.0
Fe^{2+}	(1x)	538.0	534.5	− 3.5
$O(1)^{2-}$	(1x)	538.0*	480.5	− 57.5
$O(2)^{2-}$	(1x)	452.3**	481.1	− 28.8
$O(2)^{2-}$	(1x)	476.3**	476.3	+ 24.0
Σ		2467.4	2466.9	+ 19.5 = +0.8%

* in 'FeO',   ** in Na_2O, the sequence is arbitrary

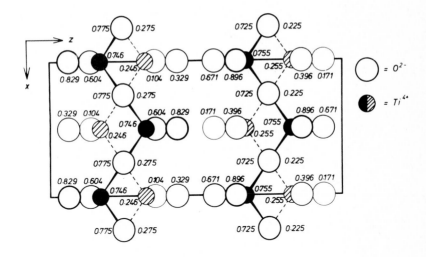

Fig. 6. $\frac{1}{\infty}[TiO_{2/1}O_{2/2}]$-chains in the structure of Rb_2TiO_3. Projection along [010].

TABLE 11

MAPLE (kcal/mol) of Rb_2TiO_3 [108]

Rb_2TiO_3	binary	ternary	$\Sigma\Delta$
$Rb(1)^+$	100.2	106.1	+ 5.9
$Rb(2)^+$	100.2	115.0	+ 14.8
Ti^{4+}	2044.9	1982.2	- 62.7
$O(1)^{2-}$	372.5[*]	518.8[***]	+146.3
$O(2)^{2-}$	610.1[**]	523.5[***]	- 86.6
$O(2)^{2-}$	610.1[**]	644.4[***]	+ 34.3
	3838.0	3890.0	+ 52.0
			(= 1.3%)

[*]Rb_2O, [**] anatase, [***] the sequence of these values is arbitrary

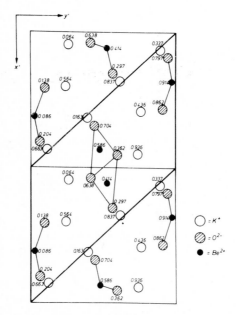

Fig. 7. Structure of K_2BeO_2; projection along [OO1].
The z-values have been indicated.

TABLE 12

MAPLE (kcal/mol) of K_2BeO_2 [113]

K_2BeO_2	binary	ternary	$\Sigma\Delta$
$K(1)^+$	104.9	103.3	− 1.9
$K(2)^+$	104.9	105.6	+ 0.7
Be^{2+}	660.8	670.7	+ 9.9
$O(1)^{2-}$	390.1[*]	562.9[***]	+172.8
$O(2)^{2-}$	660.8[**]	497.5[***]	−163.3
Σ	1921.5	1939.7	+ 18.2
			(= 0.95%)

[*] in K_2O, [**] in BeO, [***] the sequence is arbitrary

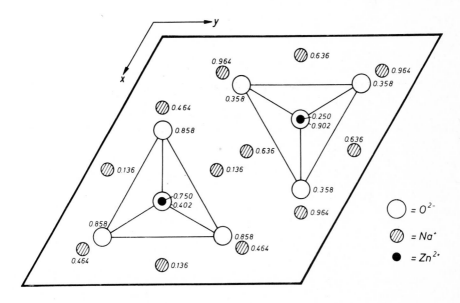

Fig. 8. Structure of Na_6ZnO_4; projection along [001].
The z-values have been indicated.

TABLE 13

MAPLE (kcal/mol) of Na_6ZnO_4 [111]

Na_6ZnO_4		binary	ternary	Δ	ΣΔ
$Na(1)^+$	3x	121.7	126.2	+ 4.5	+ 13.5
$Na(2)^+$	3x	121.7	117.7	− 4.0	− 12.0
Zn^{2+}	1x	553.6	531.8	− 21.8	− 21.8
$O(1)^{2-}$	1x	553.6*	478.7	− 74.9	− 74.9
$O(2)^{2-}$	3x	452.3**	480.3	+ 28.0	+ 84.0
Σ		3194.3	3183.1		− 11.2
Δ(%)					− 0.3
* in ZnO, ** in Na_2O.					

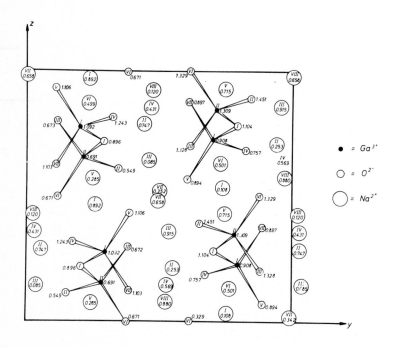

Fig. 9. Structure of $Na_8Ga_2O_7$; projection along [100].

TABLE 14
MAPLE (kcal/mol) of $Na_8Ga_2O_7$ [106]

$Na_8Ga_2O_7$	binary	ternary	$\Sigma\Delta$
$Ga(1)^{3+}$	1214.3	1170.0	− 44.3
$Ga(2)^{3+}$	1212.9	1151.9	− 61.0
$O(1)^{2-}$	604.6	573.3	− 31.3
$O(2)^{2-}$	565.3	502.6	− 62.7
$O(3)^{2-}$	584.7	515.1	− 69.6
$O(4)^{2-}$	452.3	513.2	+ 60.9
$O(5)^{2-}$	452.3	513.6	+ 61.3
$O(6)^{2-}$	452.3	518.3	+ 66.0
$O(7)^{2-}$	452.3	518.0	+ 65.7
$Na(1)^+$	121.7	117.3	− 4.4
$Na(2)^+$	121.7	128.9	+ 7.2
$Na(3)^+$	121.7	135.4	+ 13.7
$Na(4)^+$	121.7	135.6	+ 13.9
$Na(5)^+$	121.7	123.9	+ 2.2
$Na(6)^+$	121.7	126.7	+ 5.0
$Na(7)^+$	121.7	138.6	+ 16.9
$Na(8)^+$	121.7	134.7	+ 13.0
Σ	6964.6	7017.1	+ 52.5 = 0.7%

TABLE 15
MAPLE (kcal/mol) of Na_5GaO_4 [112]

Na_5GaO_4	binary	ternary	$\Sigma\Delta$
$Ga^{3+}(1/2x)$	1214.3[*]	1150.4	− 31.95
$Ga^{3+}(1/2x)$	1212.9[*]	1150.4	− 31.25
$O(1)^{2-}(1/2x)$	565.3[*]	510.5	− 27.4
$O(1)^{2-}(1/2x)$	452.3[**]	510.5	+ 29.1
$O(2)^{2-}(1/2x)$	604.6[*]	510.3	− 47.15
$O(2)^{2-}(1/2x)$	452.3[**]	510.3	+ 29.0
$O(3)^{2-}(1/2x)$	584.6[*]	514.7	− 34.95
$O(3)^{2-}(1/2x)$	452.3[**]	514.7	+ 31.2
$O(4)^{2-}$	452.3[**]	520.4	+ 68.1
$Na(1)^+$	121.7	134.4	+ 12.7
$Na(2)^+$	121.7	132.5	+ 10.8
$Na(3)^+$	121.7	119.5	− 2.2
$Na(4)^+$	121.7	121.0	− 0.7
$Na(5)^+$	121.7	124.6	+ 2.9
Σ	3830.1	3838.3	+ 8.2 = 0.2%

[*] from β-Ga_2O_3, [**] from Na_2O, the sequence is arbitrary

VII. MAPLE OF MOLECULES IN THE SOLID STATE

Taking typical molecules as ions (e.g. $Xe^{2+}F_2^-$, $Xe^{4+}F_4^-$, $Si^{4+}F_4^-$) the difference in MAPLE of the "isolated" molecule and that of the crystal is small, see Tables 16, 17, 18. But the MAPLE values of the charged particles show remarkable differences, which due to difference in sign, partly cancel each other. Due to the variation of the geometrical situation it is to expect that the "central ions" with the high values of assumed charges <u>gain</u> and the "ligands" with the small assumed charges <u>loose</u> in MAPLE

TABLE 16

MAPLE-Values of XeF_2 (kcal/mol) [114]

	XeF_2(solid)	XeF_2(gas)	Δ	Δ(%)
Σ	635.6	585.7	+ 49.9	+ 8.5
Xe^{2+}	409.8	334.7	+ 75.1	+ 22.4
F^-	112.9	125.5	- 12.6	- 10.0

TABLE 17

MAPLE-Values of XeF_4 (kcal/mol) [115]

	XeF_4(solid)	XeF_4(gas)	Δ	Δ(%)
Σ	2170.0	2087.0	+ 83	+ 4.0
Xe^{4+}	1611.2	1371.8	+239.4	+ 17.5
$F(1)^-$	140.3	178.2	- 37.9	- 21.3
$F(2)^-$	139.0	179.3	- 40.3	- 22.5

TABLE 18

MAPLE-Values of SiF_4 (kcal/mol) [116]

	SiF_4(solid)	SiF_4(gas)	Δ	Δ(%)
Σ	2680.2	2500.8	+180.2	+ 7.2
Si^{4+}	1894.2	1701.6	+192.6	+ 11.3
F^-	196.5	199.8	- 3.3	- 1.7

These values show that if the molecules under inspection would have the same amount of polarity the packing of the molecules in solid XeF_2 would be a little "better" than in solid SiF_4 and much "better" than in solid XeF_4. The values are on the other hand a warning to take MAPLE of the molecule as a whole only. A detailed discussion *must* include the Partial MAPLE

Values necessarily. So from the point of view of the "cations"
the packing is again the best in the case of XeF_2, but better
in XeF_4 than in SiF_4. The Tables 16-18 therefore show how doubtful
it is to use MAPLE as an indicator in discussing the packing of
molecules in the solid state.

This is confirmed by the MAPLE values collected in the
Tables 19-21. Here we have calculated MAPLE for hypothetical
molecules like $Cl^{4+}O_4^-$, $S^{4+}O_4^-$, $P^{4+}O_4^-$ having the crystal structure
which corresponds to the ClO_4^--, SO_4^{2-}- or PO_4^{3-}-part of the
crystal structure of $KClO_4$, $BaSO_4$ or $AlPO_4$. In view of the
fact, that the real crystal structures contain the ions K^+,
Ba^{2+} or Al^{3+} in addition, the "intermolecular" distances between
the MO_4-groups are enlarged, and correspondingly the changes
in MAPLE by condensation of the hypothetical molecules are
smaller than in the case of "undiluted" molecules like the
ones mentioned before, e.g. XeF_4.

TABLE 19
$KClO_4$, MAPLE (kcal/mol) of ClO_4-part [117]

	solid	gas	Δ	Δ %
ClO_4	2791.3	2694.6	+ 96.7	+ 3.6
Cl^{4+}	1924.2	1833.4	+ 90.8	+ 5.0
$O(1)^-$	220.9	215.3	+ 5.6	+ 2.6
$O(2)^-$	229.8	215.3	+ 14.5	− 6.7
$O(3)^-$ (2x)	208.2	215.3	− 7.1 (2x)	− 3.3

TABLE 20
$BaSO_4$, MAPLE (kcal/mol) of SO_4-part [118]

	solid	gas	Δ	Δ %
SO_4	2726.6	2634.1	+ 92.5	+ 3.5
S^{4+}	1890.0	1792.1	+ 97.9	+ 5.5
$O(1)^-$	223.6	210.5	+ 13.1	+ 6.2
$O(2)^-$	204.6	210.5	− 5.9	− 2.8
$O(3)^-$ (2x)	204.2	210.5	− 6.3	− 3.0

TABLE 21

AlPO$_4$, MAPLE (kcal/mol) of PO$_4$-part [119]

	solid	gas	Δ	Δ (%)
PO$_4$	2608.0	2500.5	+107.5	+ 4.3
P^{4+}	1842.6	1701.3	+141.3	+ 8.3
O(1)$^-$ (2x)	187.0	199.8	- 12.8	- 6.4
O(2)$^-$ (2x)	195.7	199.8	- 4.1	- 2.1

So unfortunately the unsolved problem of a precise description of packing of anions in complicated crystal structures
remains unsolved.

The same holds for the isolated groups found in the new
oxides "rich in cations" new prepared in our laboratory. Tables
22-25 give the MAPLE values of hypothetical molecules e.g. Be$_2$O$_4$
with the shape and arrangement of the (Be$_2$O$_4$)-part of the crystal
structure of K$_4$(Be$_2$O$_4$). Once again the MAPLE values are difficult to be connected precisely with the geometrical facts. In
addition these MAPLE data show that simulating vaporization
by isometric expansion of e.g. the (Be$_2$O$_4$-part of K$_4$(Be$_2$O$_4$)
with rigid molecules, the change in MAPLE as a whole is nearly
zero. In Tables 22-24 n is the factor of the stepwise multipli-
cation of the lattice constants during "vaporization" (n=1: solid
state).

TABLE 22

MAPLE (kcal/mol) of (Be$_2$O$_4$) in case of K$_4$(Be$_2$O$_4$) [113]

n	2 x Be^{2+}	2 x O(1)$^-$	2 x O(2)$^-$	Σ
1	456.8	206.2	153.6	1633.2
2	420.9	225.0	163.5	1618.8
4	414.9	228.1	166.1	1618.2
6	414.4	228.5	166.4	1618.6
8	414.1	228.6	166.5	1618.4
∞	413.9	228.6	166.5	1618.0

TABLE 23

MAPLE (kcal/mol) of $(ZnO_4)^{+O}$ in case of Na_6ZnO_4 [111]

n	1 x Zn^{4+}	1 x $O(1)^-$	3 x $O(2)^-$	Σ
1	1447.4	156.5	147.7	2047
2	1344.7	169.6	177.1	2046
4	1330.8	173.0	181.0	2047
6	1330.1	173.3	181.6	2048
8	1330.5	173.4	181.9	2049
∞	1327.9	173.6	181.5	2046

TABLE 24

MAPLE (kcal/mol) of $(FeO_3)^{+O}$ in case of $Na_4(FeO_3)$ [110]

n	Fe^{6+}	$O(1)^{2-}$	$O(2)^{2-}$	$O(3)^{2-}$	Σ
1	3466.3	552.3	566.4	553.4	5138.4
2	3217.7	624.6	658.1	631.9	5132.3
3	3192.7	632.1	667.2	640.7	5132.7
4	3185.4	633.1	699.6	642.7	5130.8
5	3183.7	633.9	670.3	643.7	5131.6
6	3182.9	634.2	670.6	644.1	5131.8
7	3182.3	634.4	670.8	644.2	5131.7
8	3181.8	634.5	670.8	644.2	5131.2
∞	3182.2	635.1	671.1	644.3	5132.7

VIII. DISCUSSION OF DETAILS OF CRYSTAL STRUCTURES

Very often a typical solid state compound, e.g. NaCl, consists of a more or less closest packing of "big anions". The holes with tetrahedral or octahedral site symmetry are partly or completely filled by "smaller cations". In such cases a description and discussion of details of the structure can be given in the usual terms of crystal chemistry.

In the case of "big cations" and/or unusual crystal structures, however, an illustrative description and/or a productive discussion of details may be very difficult, e.g. due to the fact that the C.N. of some or all of the particles can be given only in an arbitrary way. Here MAPLE is of valuable help, as may shortly be indicated in the case of Rb_2TiO_3. For a detailed discussion see [108].

TABLE 25

MAPLE (kcal/mol) of (TiO_3) in case of $Rb_2(TiO_3)$ [108]

n m m	111	122	144	188	$1 \infty \infty$	$\Delta 111/_{1 \infty \infty}$
Ti^{3+}	938	842	819	814	812	$- 126$
$O(1)^-$	149	176	184	186	188	$+ 39$
$O(2)^-$	151	182	189	191	193	$+ 42$
$O(3)^-$	219	253	261	262	265	$+ 46$
Σ	1457	1453	1453	1453	1453	

According to $Rb'_1 Rb''_1 Ti_1 O'_1 O''_1 O'''_1$ the structure contains six different particles. Let us start with the $Rb'_1 Rb''_1$-part of the structure. Calculating MAPLE for $Rb'^+_1 Rb''^-_1$ shows, that the packing of Rb' (MAPLE: 65.0 kcal/mol) and Rb" (MAPLE: 64.7 kcal/mol) is similar to that of Cu^+ and Au^- in CuAu, (both MAPLE, 69.3 kcal/mol; a-axis length chosen so that shortest distances Cu^+-Au^- are equal to Rb^+-Rb^-), in both cases both particles being geometrically "equivalent". O'" is the "bridging" O (O_B) of Rb_2TiO_3. MAPLE of the $Ti^+ O'''^-$-part shows that the packing of Ti^+ (MAPLE: 124 kcal/mol) with respect to itself is a little "better" than of O'" (MAPLE: 111 kcal/mol) in itself. MAPLE of Ti^{2+} and the "terminal" O'$^-$ and O"$^-$, that means the $Ti^{2+} O'^- O''^-$-part of the structure, shows that O'$^-$ (MAPLE: 124 kcal/mol) and O"$^-$ (MAPLE: 129 kcal/mol) adopt geometrical positions which are very similar with respect to Ti. This is confirmed by MAPLE of the $Ti^{3+} O'^- O''^- O'''^-$-part which shows that O'$^-$ (MAPLE: 149 kcal/mol) and O"$^-$ (MAPLE: 151 kcal/mol) still remain very similar as should be expected while O'"$^-$ (MAPLE: 219 kcal/mol) clearly shows the peculiarity of a bridging particle. In similar way the geometrical aspects of a structure, even in the case of a truly covalent compound like C_6H_6, can be discussed step by step in detail.

IX. MAPLE VALUES OF UNKNOWN COMPOUNDS

In case of oxocobaltates(IV) a comparison of MAPLE values of the ternary oxides with the sum of corresponding MAPLES of the binary oxides is not possible because untill now CoO_2 is unknown. Table 26 shows that even in such cases MAPLE is useful to proof the agreement of crystal structure of compounds of various compositions. Here the agreement between $K_6(Co_2O_7)$

and $Cs_2(CoO_3)$ is good. The mean value elucidated in this way
for CoO_2 is moreover in agreement with the value calculated
for a rutile-like modification of this unknown compound. But,
of course, how to prepare pure CoO_2 is still unknown.

TABLE 26
On MAPLE-Values of $K_6(Co_2O_7)$ [105] and $Cs_2(CoO_3)$ [107]

A detailed discussion is impossible with respect to the non-existence of CoO_2, but:

MAPLE $K_6Co_2O_7$:	8505 kcal/mol	
- 3 x MAPLE K_2O:	1800 kcal/mol	
MAPLE CoO_2:	6705 / 2	= 3352 kcal/mol
MAPLE Cs_2CoO_3:	3855 kcal/mol	
- MAPLE Cs_2O:	541 kcal/mol	
MAPLE CoO_2:		3314 kcal/mol

On the MAPLE-Values of 'CoO_2'
Mean Value from $K_6(Co_2O_7)$ and $Cs_2(CoO_3)$: 3333 kcal/mol

With Ti-O = 1.87 Å (in Rb_2TiO_3)
        = 1.96 Å (in rutile)
and Co-O = 1.84 Å (in Cs_2CoO_3)
we get 1.92_7Å (in 'CoO_2')
leading to
MAPLE ('CoO_2'):   3307 kcal/mol

Agreement very good. Thus: CoO_2 may be a rutile

X. CLOSING REMARK

To deal with crystal structures without using models is
still an unsolved problem. MAPLE is a good tool as has been
shown here by some examples out of a real lot of compounds we
have studied by MAPLE more intensively during the last   years.

In solving new crystal structures MAPLE too is a good
guide. This holds especially in case of compounds where the
distribution of small particles like Li^+ on holes which are in
excess is difficult to be solved by X-Ray studies. We have
studied this e.g. in the case of compounds like Li_3InO_3 and
Li_5GaO_4 [120]. But as long as the crystal structure of new

156   , R. Hoppe

compounds can not predicted exactly, most of the above
mentioned problems remain unresolved.

ACKNOWLEDGEMENTS

I thank the Deutsche Forschungsgemeinschaft for
valuable financial support. In addition I thank D. Wald for
help in doing the numerous calculations and for his critical
view.

References

[1] R. Hoppe, Angew.Chem. 78 (1966) 52; Angew.Chem.Intern. Ed. 5 (1966) 95.

[2] R. Hoppe, Angew.Chem. 82 (1970) 7; Angew.Chem.Intern. Ed. 9 (1970) 25.

[3] R. Hoppe, Madelung constants as a new guide to the structural chemistry of solids in: Advances in Fluorine Chemistry, Vol. 6 (Butterworths, London, 1970) p. 387.

[4] R. Hoppe, Der Madelunganteil der Gitterenergie, MAPLE, als Strukturcharakteristikum in der Festkörperchemie; Vorträge aus dem Gebiet der Festkörperchemie, Klemm-Festschrift (Münster 1971).

[5] R. Hoppe, Izv. Jugoslav. Centr. Krist. (Zagreb) 8 (1973) 21.

[6] M. Barlow and Ch.C. Meredith, Z. Kristallogr., Kristall-geometr., Kristallphysik, Kristallchem. 130 (1969) 304.

[7] C. Caranoni, R. Favier, L. Capella and A. Tranquard, Compt. Rend. (Paris) C270 (1970) 1795.

[8] R.W.G. Wyckoff, Crystal Structures, Vol. 1 (Wiley, New York, 1963) p. 300.

[9] Ibidem, Vol. 1, p. 242.

[10] Ibidem, Vol. 1, p. 175.

[11] Ibidem, Vol. 1, p. 177.

[12] Ibidem, Vol. 1, p. 310.

[13] J.H. Burns, J.R. Peterson and R.D. Baybarz, J.Inorg.Nucl.Chem. 35 (1973) 1171.

[14] B. Morosin and A. Narath, J.Chem.Phys. 40 (1964) 1958.

[15] A.J.C. Wilson, Structure Reports, Vol. 9, 1942-44 (Oosthoek, Utrecht, 1955) p. 159.

[16] J.C. Taylor and P.W. Wilson, Acta Cryst. B30 (1974) 1216.

[17] H. Bärnighausen and N. Schultz, Acta Cryst. B25 (1969) 1104.

[18] H. Bärnighausen, personal communication.

[19] G.A. Jeffrey and M. Vlasse, Inorg.Chem. 6 (1967) 396.

[20] G. Thiele, K. Brodersen, E. Kruse and B. Holle, Natur-wissenschaften 54 (1967) 615.

[21] G. Thiele, K. Brodersen, E. Kruse and B. Holle, Chem.Ber. 101 (1968) 2771.

[22] D.K. Smith, H.W. Newkirk and J.S. Kahn, J. Electrochem. Soc. 111 (1964) 78.

[23] D.K. Smith and C.F. Cline, Acta Cryst. 18 (1965) 393.

[24] G. Malmros, Acta Chim. Scand. 24 (1970) 384.

[25] B. Aurivillius and G. Malmros, Kungl. Tekn. Högskolans Handl. 291 (1972) 544.

[26] G.S. Smith and P.B. Isaacs, Acta Cryst. 17 (1964) 842.

[27] K.J. Seifert and H. Nowotny, Monatsh. Chem. 102 (1971) 1006.

[28] W.H. Baur and A.A. Khan, Acta Cryst. B27 (1971) 2133.

[29] See Ref. [8], Vol. 1, p. 99.

[30] Ibidem, Vol. 1, p. 98.

[31] J. Leciejewicz, Acta Cryst. 14 (1961) 1304.

[32] Ibidem, 14 (1961) 66.

[33] B.M. Gatehouse and A.D. Wadsley, Acta Cryst. 17 (1964) 1545.

[34] S. Andersson, Z. Anorg. Allg. Chem. 351 (1967) 106.

[35] W. Mertin, S. Andersson and R. Gruehn, J. Solid State Chem. 1 (1970) 419.

[36] F. Laves, W. Petter and H. Wulf, Naturwissenschaften 51 (1964) 633.

[37] R. Gruehn, J. Less-Common Metals 11 (1966) 119.

[38] J.D. Donaldson, W. Moser and W.B. Simpson, Acta Cryst. A16 (1963) 22.

[39] See Ref. [15], Vol. 11, 1947-48 (Oosthoek, Utrecht, 1951) p. 238.

[40] See Ref. [8], Vol. 1, p. 392.

[41] O. Muller and R. Roy, J. Less-Common Metals 16 (1968) 129.

[42] See Ref. [8], Vol. 1, p. 312.

[43] Ibidem, Vol. 1, p. 313.

[44] Ibidem, Vol. 1, p. 318.

[45] W.A. Dollase, Z. Kristallogr., Kristallgeometr., Kristall-physik, Kristallchem. 121 (1965) 369.

[46] W.A. Dollase, Acta Cryst. 23 (1967) 617.

[47] See Ref. [8], Vol. 1, p. 315.

[48] J. Shropshire, P.P. Keat and P.A. Vaughan, Z. Kristallogr., Kristallgeometr., Kristallphysik, Kristallchem. 112 (1959) 409.

[49] W.H. Baur and A.A. Khan, Acta Cryst. B27 (1971) 2133.

[50] T. Araki and T. Zoltai, Z. Kristallogr., Kristallgeometr., Kristallphysik, Kristallchem. 129 (1969) 381.

[51] H. Beyer, Z. Kristallogr., Kristallgeometr., Kristallphysik, Kristallchem. 124 (1967) 228.

[52] O. Lindqvist, Acta Chim.Scand. 22 (1968) 977.

[53] S.C. Abrahams and J.L. Bernstein, J. Chem. Phys. 55 (1971) 3206.

[54] See Ref. [8], Vol. 1, p. 254.

[55] Ibidem, Vol. 1, p. 253.

[56] P.Y. Simons and F. Dachille, Acta Cryst. 23 (1967) 334.

[57] J.M.D. Coey, Acta Cryst. B26 (1970) 1876.

[58] R.D. Shannon and C.T. Prewitt, J. Solid State Chem. 2 (1970) 134.

[59] J.W. Biesterbos and J. Hornstra, J. Less-Common Metals 30 (1973) 121.

[60] R.W.G. Wyckoff, Crystal Structures, Vol. 2 (Wiley, New York, 1964) p. 95.

[61] Ibidem, Vol. 2, p. 17.

[62] C. Svensson, Acta Crysta. B30 (1974) 458.

[63] W.B. Pearson, Structure Reports, Vol. 17, 1953 (Oosthoek, Utrecht, 1963) p. 396.

[64] B.O. Loopstra and H.M. Rietveld, Acta Cryst. B25 (1969) 1420.

[65] W.L. Kehl, R.G. Hay and D. Wahl, J. Appl. Phys. 23 (1952) 212.

[66] See Ref. [15], Vol. 8, 1940-41 (Oosthoek, Utrecht, 1956) p. 143.

[67] Ibidem, Vol. 12, 1949 (Oosthoek, Utrecht, 1952) p. 182.

[68] Ibidem, Vol. 8, 1940-41, (Oosthoek, Utrecht, 1956) p. 145.

[69] H.T. Evans, jr. Nature 232 (1971) 69.

[70] See Ref. [8], Vol. 1, p. 333.

[71] J.T. Mason, M.C. Jha, D.M. Bailey and P. Chiotti, J. Less-Common Metals 35 (1974) 331.

[72] B.Brown, T.L. Hall and P.T. Moseley, J. Chem. Soc. Dalton Trans. (1973) 686.

[73] K.A. Becker, K. Plieth and J.N. Stranski, Progr. Inorg. Chem., 4 (1962) 1.

[74] G.E. Gurr, P.W. Montgomery, C.D. Knutson and B.T. Gorres, Acta Cryst. B26 (1970) 906.

[75] C.T. Prewitt and R.D. Shannon, Acta Cryst. B24 (1968) 869.

[76] See Ref. [8], Vol. 1, p. 86.

[77] H.T. Evans, Jr., Nature 232 (1971) 69.

[78] J.E. Iglesias, K.E. Pachali and H. Steinfink, Mater. Res. Bull. 7 (1972) 1247.

[79] See Ref. [8], Vol. 1, p. 242.

[80] E. Hubbert-Paletta and H.K. Müller-Buschbaum, Z. Anorg. Allg. Chem., 363 (1968) 145.

[81] M. Gonrand, Bull. Soc. Franç. Minéral. Crist. 96 (1973) 166.

[82] See Ref. [8], Vol. 1, p. 89.

[83] M.A. Hepworth, K.H. Jack, R.D. Peacock and G.J. Westland,
     Acta Cryst. 10 (1957) 63.

[84] D. Babel, F. Wall and G. Heyer, Z. Naturforsch. 29b (1974)
     139.

[85] A. Tressaud, J. Graly and J. Portier, Bull.Soc.Franç.Minéral.
     Crist., 92 (1969) 335.

[86] D. Schmitz and W. Bronger, Z. Anorg. Allg. Chem. 402 (1973)
     225.

[87] J.E. Iglesias, K.E. Pachali and H. Steinfink, J. Solid
     State Chem. 9 (1974) 6.

[88] C. Billy and H.M. Haendler, J. Am. Chem. Soc. 79 (1957) 1049.

[89] H.G. von Schnering, Z. Anorg. Allg. Chem. 400 (1973) 201.

[90] K. Knox, Acta Cryst. 13 (1960) 507.

[91] Kang Kun Wu and J.D. Brown, Mater. Res. Bull. 8 (1973) 593.

[92] A.J. Edwards and P. Taylor, Chem. Comm. (1971) 1376.

[93] A.J. Edwards, personal communication (1971).

[94] J.H. Burns, Acta Cryst. 15 (1962) 1098.

[95] See Ref. [60], Vol. 2, p. 405.

[96] Ibidem, Vol. 2, p. 414.

[97] H.G. von Schnering, R. Hoppe and J. Zemann, Z. Anorg. Chem.
     305 (1960) 17.

[98] J.A. Bland, Acta Cryst. 14 (1961) 875.

[99] K.K. Wu and I.D. Brown, Acta Cryst. B29 (1973) 2009.

[100] H. Sabrowsky and R. Hoppe, Z. Anorg. Allg. Chem. 358 (1968)
      241.

[101] H. Rieck and R. Hoppe, Z. Anorg. Allg. Chem. 400 (1973) 311.

[102] H. Rieck and R. Hoppe, Z. Anorg. Allg. Chem. 392 (1972) 193.

[103] M. Jansen and R. Hoppe, Z. Anorg. Allg. Chem. 398 (1973) 54.

[104] M. Jansen and R. Hoppe, Z. Anorg. Allg. Chem., to be
      published.

[105] M. Jansen and R. Hoppe, Naturwissenschaften 60 (1973) 104.

[106] D. Fink and R. Hoppe, Z. Anorg. Allg. Chem., to be published.

[107] M. Jansen and R. Hoppe, Z. Anorg. Allg. Chem. 408 (1974) 75.

[108] W. Schartau and R. Hoppe, Z. Anorg. Allg. Chem. 408 (1974) 60.

[109] H. Rieck and R. Hoppe, Z. Anorg. Allg. Chem. 408 (1974) 151.

[110] H. Rieck and R. Hoppe, Naturwissenschaften 61 (1974) 126.

[111] P. Kastner and R. Hoppe, Z. Anorg. Allg. Chem., to be
      published.

[112] D. Fink and R. Hoppe, Z. Anorg. Allg. Chem., to be published.

[113] P. Kastner and R. Hoppe, Naturwissenschaften 61 (1974) 79.

[114] H.A. Levy and P.A. Argon, J. Am. Chem. Soc. 85 (1963) 241.

[115] R. Hoppe, Fortschr. Chem. Forsch. 5 (1965) 243.

[116] See Ref. [60], Vol. 2, p. 127.

[117] N.V. Mani, Proc. Indian Acad. Sci. A46 (1957) 143.

[118] K. Sahl, Beitr. Mineral. Petrogr. 9 (1963) 111.

[119] R.W.G. Wyckoff, Crystal Structures, Vol. 3 (Wiley, New York, 1965) p. 30.

[120] R. Hoppe and F. Stewner, Acta Cryst. B27 (1971) 621.

Crystal Structure and Chemical Bonding in Inorganic Chemistry
Eds. C.J.M. Rooymans and A. Rabenau
© 1975, North-Holland Publishing Company, The Netherlands

SIMPLE RULES FOR ALLOYING

A.R. Miedema

Philips Research Laboratories
Eindhoven, The Netherlands

and

R. Boom[*]) and F.R. de Boer

Natuurkundig Laboratorium,
Universiteit van Amsterdam, The Netherlands

SUMMARY

The heat of formation of alloys is described in terms of a simple atomic model. An alloy is suggested to be constructed from Wigner-Seitz atomic cells taken from the pure metallic elements. The alloying energy then originates from a change in boundary conditions when dissimilar atoms get into contact.

There are two terms in the heat of formation, ΔH, a negative one arising from removing the difference in chemical potential between the two types of atomic cells and a positive term that reflects the original discontinuity in the density of electrons at the boundary between dissimilar atomic cells.

The model accounts for the heat of mixing of liquid metals as well as for the heats of formation of binary intermetallic compounds with one or two transition metals. The results are formulated in terms of simple rules for alloy formation. It is furthermore demonstrated that the description (and hence the predictions) of ΔH for alloys of transition metals is sufficiently accurate to be of practical interest.

The present model conflicts with descriptions of heats of formation of transition metal alloys in terms of the Engel-Brewer theory.

I. INTRODUCTION

The chemistry of alloys offers an interesting problem. In contrast to insulating inorganic compounds, for which either a description in terms of ions and Madelung energies [1] or one [2]

---

[*])Present address: Research Laboratories Hoogovens, IJmuiden B.V. IJmuiden, The Netherlands.

by means of shared electron pairs can be useful to predict
stabilities, one had until recently no comparable description of
the factors that determine the stabilities of alloys, relative
to that of the pure metals. One might expect that ionicity, as
induced by the difference in electronegativity between the atoms,
also plays a role in alloys. However, since the "ionic charge"
in metallic alloys can have any value (including non-integral
ones), the Madelung sum approach is not straight-forward, while
the concept of shared electron pairs too is hard to apply for
metals. In addition there must be large energy effects, other
than ionic ones, since otherwise it is impossible to understand
why there are so many combinations of two metals that refuse to
form alloys at any composition and even show immiscibility in the
liquid state.

In a series of papers [3,4] we recently have introduced
a simple atomic model that accounts for energy effects in alloys.
The model is in principle empirical but since the first idea
was published it gradually has got a physical basis, although
there are a number of questions that still have to be solved yet.
In this paper we will take once more the empirical approach; we
describe the model and demonstrate its validity. In order to
avoid ample discussions of the more sophisticated questions that
arise we refer, where possible, to versions of this model
published elsewhere. Our main concern will be to convince the
reader of the usefulness of the concept of individual atoms for
metallic alloys and to make clear that one can predict the
enthalpies of formation of alloys with an accuracy that is
comparable to that of the experimental values.

II. THE MODEL

The basic assumption, we start from, is that the Wigner-Seitz
concept of atomic cells, that has been used extensively in
theoretical descriptions of pure metals [5], is still meaningful
for the two types of cells in a binary alloy. It is assumed that
in first approximation atomic cells of the two metals A and B
are similar in the alloy and in the pure metal, i.e. the atomic
volumes are taken the same. One should note that atoms not
necessarily have to be spheres; an assembly of atoms has to fill
the whole space which in a two-dimensional drawing makes them
squares for pure metals (fig. 1a). In contructing the alloy the

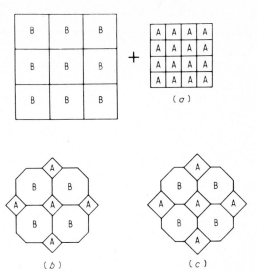

Fig. 1. The atomic model for alloys. Wigner-Seitz atomic cells are
taken from the pure metals to form an alloy. The heat of
formation is derived from the change in boundary conditions
at the positions where there are contacts between dissimilar cells. One effect is charge transfer, that will
influence the cell volumes, indicated at b and c.

B cells have been changed in shape in order to be able to obtain
a good packing. For pure metals it is known that a change of the
shape of a Wigner-Seitz cell takes only little energy as long as
its volume remains the same (allotropy). If fig. 1b would cover
the full story, an alloy would be equivalent to a mechanical
mixture of metals A and B and the heat of formation of the alloy
would be zero. However, the boundary conditions have become
different at those points where cells A and B are in contact.
We shall derive the heat of formation of alloys directly from
this change in boundary conditions.

This is an important advantage above the treatments of the
heats of formation of chemical compounds in which a type of Born-
Haber cycle is used, where one first evaporates atoms from the
pure solids and later gains energies such as the Madelung energy
of the formed ionic compounds. In general two large energy contributions have then to be subtracted, so that heats of formation
of a few kcal/g-at only are difficult to reproduce. In our case,
however, we start from a mechanical mixture of already metallic
atoms and the energy effect to be derived is directly the heat
of formation.

The change in boundary conditions when transferring metal
atoms from pure metals A and B to the alloy AB implies two
different energy effects. The first is a discontinuity in the
density of electrons (n_{ws}) at the boundary between dissimilar
Wigner-Seitz cells. This density discontinuity, Δn_{ws}, has to
be smoothed. Originally the atomic volume and the corresponding
electron density in between the atoms is such that there is an
energy minimum for each of the two metals. Hence any change in
n_{ws} will lead to a positive contribution to the energy of
alloying. This positive energy contribution can be expected
to be proportional to $(\Delta n_{ws})^2$ in first approximation.

Secondly there will be a difference in the chemical
potential for electrons in the two types of pure metal cells
(another way of saying is that there is a difference in electronegativity for the two elements in the metallic state or, alternatively, that the Fermi energies as measured relative to vacuum
are different for the two metals). The chemical potential, which
we denote by ϕ^*, cannot be allowed to vary in an alloy. Hence
there will be some charge redistribution which results in a
negative contribution to the heat of formation of the alloy. In
first approximation the energy effect will be proportional to
$(\Delta\phi^*)^2$. Adding the two energy contributions gives:

$$\Delta H \sim [-Pe(\Delta\phi^*)^2 + Q(\Delta n_{ws})^2] \tag{1}$$

(e is the elementary charge).

It is possible to look upon the first term as if it was
the energy of an electric dipole layer that is generated at
tne contact surface between different metals. As a matter of fact
the energy of the dipole layer is proportional to the total sur-
face area S, it is proportional to $(\Delta\phi_{exp})^2$, where $\Delta\phi_{exp}$ is
the difference in work function for the two metals, and it is
inversely proportional to a width parameter that is related to
the electrostatic shielding length. The total surface area is,
apart from a function of the concentrations of the two metals,
proportional to $\overline{V}_m{}^{2/3}$ where V_m is the molar volume. It turns
out that in spite of this P is an approximate constant for
alloys of widely different metals, due to the fact that the
dependence on $V_m{}^{2/3}$ and that on the electrostatic shielding length
(read: density of electrons) in practice nearly cancel.

From physical considerations of the same kind it can be
argued that the constant Q in relation (1) in fact contains
a factor $(n_{ws})^{-4/3}$ so that relation (1) preferably is rewritten as

$$\Delta H \sim [-Pe(\Delta\phi^{*})^2 + Q'n_{ws}^{-4/3}(\Delta n_{ws})^2] \qquad (2)$$

which is equivalent to

$$\Delta H \sim [-Pe(\Delta\phi^{*})^2 + Q_o(\Delta n_{ws}^{1/3})^2] \qquad (3)$$

Before being able to compare relation (3) with experimental
data we have to answer three questions:
a) which value has ϕ^{*} for metallic elements?
b) what is $n_{ws}^{1/3}$ for pure metals?
c) where do we find a fairly complete set of experimental values
   for ΔH?

In fact the answer to the first question has been given
already. Treating atoms as pieces of metals that have a resem-
blance to macroscopic metals one identifies ϕ^{*} with ϕ_{exp}. Hence
we will try to use experimental values of the work function as
a measure for ϕ^{*}.

The second question offers no problem for simple metals
i.e. the non-transition metals. For these elements one can derive
n_{ws} by calculating the charge distribution in a metallic crystal
as if it was a superposition of the charge distributions for
free atoms, that occupy lattice positions. This theoretically
estimated value of n_{ws} depends little on the actual crystal
structure as long as the atomic volume is kept the same. Unfortunately one cannot rely on a calculation of this type in
case of transition metals since here metallic and atomic charge
distributions can be widely different. However, there is another
way to estimate n_{ws} for transition metals. We make use of the
fact that there is a simple relation between the electron density
between atoms and the compressibility of a metal. The existence
of such a relation is not surprising. If the density has a low
value, one expects that it is easier to compress the metal than
in case n_{ws} has a high value. As shown in fig. 2 for all non-
transition metals, the theoretically derived values for n_{ws} are
related to the bulk modulus B and the molar volume V_m of the
pure metals by

$$n_{ws}^2 \sim B/V_m \qquad (4)$$

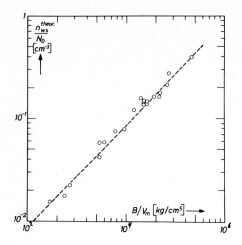

<u>Fig. 2.</u> The relation between the electron density at the boundary
         of the Wigner-Seitz atomic cell and the ratio of the bulk
         modulus, B, and the molar volume, V_m, for pure non-
         transition metals. Values for B and V_m are from
         Gschneidner [6].

     We simply decide to use this relation between electron
density and bulk modulus, which is very convincing for non-
transition metals, to estimate n_{ws} for transition metals.
     The last question concerns ΔH. It was possible to find in
the literature a representative set of experimental values of
the heat of mixing for liquid alloys of non-transition metals,
so we have no difficulties in comparing model and experiment for
this group of metals. However, one would like to do more and
consider for instance solid alloys of transition metals too. Here,
however, the existing thermodynamic information is quite limited.
We can still perform the analysis if we do not consider absolute
values of ΔH but first focus on its sign. Only the sign of ΔH
is known for the majority of binary solid alloy systems. If, in
a binary system, there are intermetallic compounds that are stable
at low temperatures one can be quite sure that their heat of
formation is negative. If, on the contrary, there are no compounds
in a system while at the same time both mutual solubilities are
small (say < 10 at %) one can conclude that ΔH is positive.
     When analyzing information on the sign of ΔH we do not
need to know its concentration dependence. From relation (3) one

immediately derives that $\Delta H=0$ corresponds to

$$|\Delta \phi^* / (\Delta n_{ws}^{1/3})| \quad = (Q_o/Pe)^{\frac{1}{2}}$$

Hence the ratio $|\Delta \phi^* / \Delta n_{ws}^{1/3}|$ is predicted to characterize the sign of ΔH.

III. ANALYSIS OF THE SIGN OF ΔH

We start with liquid alloys of two non-transition metals. In fig. 3 the sign of ΔH is plotted as a function of the difference in the experimental values of the work function and the theoretical values of n_{ws} for the two metals. Each data points represents a binary system. In order to reduce the influence of experimental uncertainties a positive sign indicates that the integral heat of

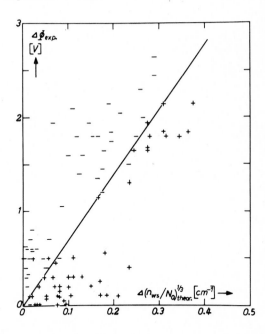

Fig. 3. The sign of the heat of mixing of liquid alloys of two non-transition metals, plotted as a function of $\Delta \phi_{exp}$ and $\Delta(n_{ws}/N_o)^{1/3}$. The heat of mixing at the 50/50 composition is given as - for $\Delta H < -600$ cal/g-at;
    + for $\Delta H > +600$ cal/g-at or in case of liquid immiscibility. Values for ϕ_{exp} are from [7] and [8]. For references to the experimental ΔH data see [9] and [10]. N_o is the Avogadro's number.

mixing of the equiatomic alloy exceeds +600 cal/g-at. Consequently
a minus sign indicates ΔH < -600 cal/g-at. In addition we have
included with a positive sign systems for which liquid immiscibility has been reported.

One may see that relation (3) is followed closely. By means
of a straight line through the origin the region that contains
predominantly plus signs can be separated from the region
containing mainly minus signs. From the slope of the line the
value of $(Q_0/Pe)^{\frac{1}{2}}$ can be derived.

We have investigated to which extent the correlation can
be increases by allowing $\emptyset^*$ to be somewhat different from $\emptyset_{exp}$,
and n_{ws} to be somewhat different from its theoretical estimate.
Upon changing a parameter for a given element, all data points
for binary alloys containing that element will make a shift
in fig. 3. By trial and error and permitting only small deviations
from $\emptyset_{exp}$ and n_{ws} (theoretical) we obtain the result of fig. 4.

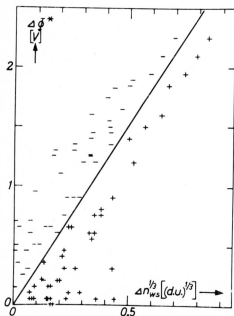

Fig. 4. The sign of the heat of mixing of liquid alloys of two
non-transition metals (see table I) plotted as a function
of $\Delta\emptyset^*$ and $\Delta n_{ws}^{1/3}$. The parameters $\emptyset^*$ and n_{ws} are per-
mitted to be somewhat different from their experimental
and theoretical values, respectively. Signs - and + as in
fig. 3.

TABLE 1

The parameters needed in the relations (3), (5) or (7) for the heats of formation of alloys.

ϕ^* represents the chemical potential for electrons (the electronegativity parameter), n_{ws} is the density of electrons at the boundary of the Wigner-Seitz atomic cell and V_m is the molar volume. The units of n_{ws} are such that the density is near 1 for Li; the density unit is derived from $(B/V_m)^{\frac{1}{2}} = 10^2 \ \text{kg}^{\frac{1}{2}}\text{cm}^{-5/2}$.

Transition metals	ϕ^* (V)	$n_{ws}^{1/3}$	$V_m^{2/3}$ (cm^2)	Non-trans. metals	ϕ^* (Volt)	$n_{ws}^{1/3}$	$V_m^{2/3}$ (cm^2)
Sc	3.25	1.27	6.1	Li	2.85	0.98	5.5
Ti	3.65	1.47	4.8	Na	2.70	0.82	8.3
V	4.25	1.64	4.1	K	2.25	0.65	12.8
Cr	4.65	1.73	3.7	Rb	2.10	0.60	14.6
Mn	4.45	1.61	3.8	Cs	1.95	0.55	16.8
Fe	4.93	1.77	3.7	Cu	4.55	1.47	3.7
Co	5.10	1.75	3.5	Ag	4.45	1.39	4.8
Ni	5.20	1.75	3.5	Au	5.15	1.57	4.8
Y**)	3.20	1.21	7.3	Ca	2.55	0.91	8.8
Zr	3.40	1.39	5.8	Sr	2.40	0.84	10.2
Nb	4.00	1.62	4.9	Ba	2.32	0.81	11.3
Mo	4.65	1.77	4.4	Be	4.20	1.60	2.9
Tc	5.30	1.81	4.2	Mg	3.45	1.17	5.8
Ru	5.55	1.87	4.1	Zn	4.10	1.32	4.4
Rh	5.40	1.76	4.1	Cd	4.05	1.24	5.5
Pd	5.60	1.65	4.3	Hg	4.20	1.24	5.8
La	3.05	1.09	8.0	Al	4.20	1.39	4.6
Hf	3.55	1.43	5.6	Ga	4.10	1.31	5.2*)
Ta	4.05	1.63	4.9	In	3.90	1.17	6.3
W	4.80	1.81	4.5	Tl	3.90	1.12	6.6
Re	5.50	1.90	4.3	Si	4.70	1.50	4.2x)
Os	5.55	1.89	4.2	Ge	4.55	1.37	4.6x)
Ir	5.55	1.83	4.2	Sn	4.15	1.24	6.4
Pt	5.65	1.78	4.4	Pb	4.10	1.15	6.9
Th	3.30	1.28	7.3	As	4.80	1.40	5.2x)
U	4.05	1.56	5.6	Sb	4.40	1.26	6.6x)
Pu	3.80	1.44	5.2	Bi	4.15	1.16	7.2x)

*) Volume reduced for unusual crystal structure i.e. the fact that the metal contracts upon melting.

**) For the rare earth metals with the smaller molar volumes the values for yttrium can be used.

There is only one minus sign in the positive region, Al-Sb.
Fig. 4 contains more data points than fig. 3, since we have
here also included elements like Si and Ge for which $\emptyset_{exp}$ and
n_{ws} are difficult to obtain as the pure elements are not in an
ordinary metallic state. Here $\emptyset^*$ and n_{ws} are defined at
"reasonable values", see table 1. In fig. 5 we demonstrate that
the restriction to systems for which $|\Delta H| > 600$ cal/g-at is not
essential. The open points represent liquid alloys for which
$-600 < \Delta H < 0$ cal/g-at, the majority of filled circles lie below
the straight line $\Delta H = 0$ taken from fig. 4. Also, it is striking
that the data points are concentrated near the origin. This
is in good agreement with relation (3) since in a quadratic
relation between ΔH and $\Delta\emptyset^*$, $\Delta n_{ws}^{1/3}$, ΔH values will be small
near the origin $\Delta\emptyset^* = 0$, $\Delta n_{ws}^{1/3} = 0$.

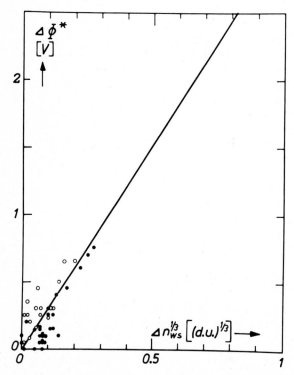

Fig. 5. The values of $\Delta\emptyset^*$ and $\Delta n_{ws}^{1/3}$ for those liquid alloys of
two non-transition metals for which ΔH has been measured
to be smaller than 600 cal/g-at.
(o) $-600 < \Delta H < 0$;

(•) $0 < \Delta H < +600$ cal/g-at.

In fig. 6 we analyse solid alloys of two transition metals and solid alloys in which one metal is a transition metal and the other one an alkali metal, an alkaline-earth metal or Cu, Ag, Au. We include 27 transition metals (see table 1) so that the total number of binary systems considered here is $27(26/2+11)=648$. In fig. 6 the sign of ΔH for a given binary system is determined from the occurrence or non-occurrence of intermetallic compounds, as discussed above. The parameters are ϕ_{exp} and n_{ws} as determined by means of relation (4) from experimental values of ϕ and B/V_m of pure metals. We include those transition metals for which recently experiments [11] on the work function of polycrystalline samples have been reported. Information on phase diagrams is taken from the

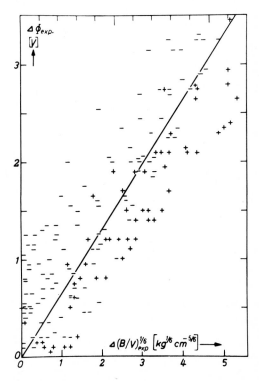

Fig. 6. The sign of the heat of formation, ΔH, for solid alloys of two transition metals or of one transiton metal and one of the metals Li, Na, K, Rb, Cs, Cu, Ag, Au, Ca, Sr, Ba. The sign of ΔH is plotted as a function of $\Delta \phi_{exp}$ and of the difference in $n_{ws}^{1/3}$ as derived from experimental values of the bulk modulus B and the molar volume V_m for pure metals.

Handbooks by Hansen and Anderko, Elliot and Shunk [12], supplemented by recent literature data.

Fig. 7 is derived from fig. 6 by taking some freedom to
choose values for $\emptyset^*$ and n_{ws} different from their experimental
values. It is shown that with $\emptyset^*$ and n_{ws} as collected in table 1

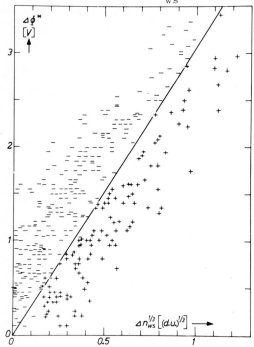

Fig. 7. The sign of ΔH for solid alloys of the same group of
metals as plotted in fig. 6 here as a function of $\Delta \emptyset^*$
and $\Delta n_{ws}^{1/3}$. For $\emptyset^*$ and $n_{ws}^{1/3}$ the preferred values, collected in table 1, have been used.

a nearly perfect agreement with relation (3) is obtained. There
are only two plus signs in the negative region while there are
four minus signs in the positive region. All exceptions lie
very near the straight line that represents $\Delta H=0$, so that they
are not disturbing. Note that relation (3) accounts for at first
sight surprising experimental results as a difference in
alloying behaviour of Au and Ag with respect to V, Cr, Co, Nb, Ta
and U (Au forms compounds while Ag does not give solid alloys).
It furthermore shows why Ni forms compounds with Ca and Sr and
not with Ba, that Pt alloys with Li and Na but not with K while
it is also shown why among the rare earth metals there is a

a difference in chemical behaviour with respect to Fe (or Re, or Mn) (the rare earth metals with the smaller molar volumes are represented by the value for yttrium in Table 1.

In fig. 8 we have included systems belonging to the above group of 648 for which there is no experimental information on the phase diagram (crosses) as well as systems for which one would expect ΔH to be near zero (circles), since there are no compounds

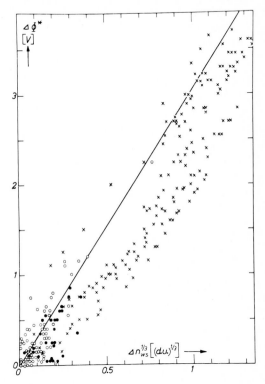

Fig. 8. The values of $\Delta\phi^*$ and $\Delta n_{ws}^{1/3}$ for binary alloys of transition metals (group of figs. 6 and 7) for which there is no experimental information (crosses) or for which ΔH is expected to be near zero.
(o) no intermetallic compounds and at least one of the solid solubilities is larger than 10 at%;

(•) same as open circles but either the solubility decreases to low values at low temperature or there is a miscibility gap in the solid state, while the pure metals have the same crystal structure.

while at least one of the solid solubilities is larger than 10 at%.
Filled circles indicate that solid solubilities become small at
low temperatures or represent a binary system in which the pure
components have the same crystal structure but nevertheless show
a miscibility gap in the solid state. One can see that again
the circles tend to be near to the origin, as expected. Also,
the filled circles that might represent systems for which ΔH is
small and positive, mainly lie below the line $\Delta H=0$. In fact it
is a strong point in favour of the present model that it accounts
for the presence of a miscibility gap in such systems as Ir-Pd
and Rh-Pd, where the pure metals are both face-centered-cubic and
differ only little in atomic volume.

The majority of uninvestigated systems lies in the region
where we predict ΔH to be positive, which indicates that the
metallurgists who decided not to investigate these combinations
have had a sound intuition. Nevertheless there are a few (i.e. 13)
uninvestigated binary systems for which we predict stable alloys
to exist. These are Mn-Os, Ni-Ir, Tc with La, Th, U, or Pu; Pd with
Na, K, Rb or Cs, Ca-Ru and Ir with Li or Ba.

As discussed above the slope of the straight line drawn in
the figs. 3-8 corresponds to the ratio of the two empirical constants Q_o/P in relation (3). It has to be emphasized that in all
figures discussed thus far the straight line drawn has the same
slope. In other words the ratio Q_o/P is approximately a "universal"
constant, that applies to solid alloys of transition metals as well
as to liquid alloys of non-transition metals.

We can also make the $\Delta\phi^*$, $\Delta n_{ws}^{1/3}$ plot for the heat of mixing
of liquid alloys of transition metals (the group of metals of
fig. 6). Here the information is limited in the sense that there
are practically no systems for which ΔH has been measured calori-
metrically. All data on liquid inmiscibility (ΔH-positive) agree
with relation (3) and the slope $(Q_o/Pe)^{\frac{1}{2}}$ one finds [10] is the
same as for all other types of alloys.

The same plot but now for solid alloys of two non-transition
metals is a less convincing one. The reason for this is that the
present description of the heats of formation does not depend
on crystal structure apart from the fact that in ordered alloys
the contact area between dissimilar atoms is larger than in
solutions. Among the solid compounds of two non-transition metals
we meet the well-known $A^N B^{8-N}$ semiconductors. The crystal structure

that makes these compounds semiconductors has a relatively large influence on the heat of formation of the compounds too. These Brillouin-zone type energy contributions are not included in relation (3) so that in order to keep things simple we leave out solid alloys of two non-transition metals in this article.

At the same time we can conclude that for solid alloys of two transition metals crystal structure-determined energy contributions play a minor role. The main terms in the alloy energy are similar for liquid and solid transition metal alloys, apart from the fact that the number of dissimilar neighbours can be larger in a solid compound. Structure-dependent energy contributions that at one hand are large enough to determine the stable phases and their crystal structure are on the other hand small relative to the energy contributions included in relation (3), at least for transition metal alloys.

IV. ALLOYS OF d-METALS AND p-METALS

A group of binary alloys not mentioned so far is that in which the first metal is one of the 27 transition metals of table 1 and the other metal is a polyvalent non-transition metal, for instance Al, Ga, In, Tl (trivalent), Sn, Pb (tetravalent) or Sb and Bi (pentavalent). In fig. 9 we have analysed the sign of ΔH for compounds of this type. A minus sign again indicates that there are stable compounds in a system, a positive sign means that there are no compounds. The situation of no compounds and a solid solubility larger than 10 at% does not occur. We first use again experimental values for $\emptyset$ and $n_{ws}^{1/3}$ as derived from experimental values of B/V_m, as before. One may deduce from fig. 9 that the parameters $\emptyset$ and $n_{ws}^{1/3}$ are useful in a description of the sign of ΔH for this group of alloys too. However, ΔH is not described by relation (3); no straight line through the origin can be drawn that separates the plus and minus regions. The demarcation line of fig. 9 is a hyperbola, which can be described by adding a negative constant in relation (3):

$$\Delta H \sim -Pe\left(\Delta\emptyset^{*}\right)^{2} + Q_{0}\left(\Delta n_{ws}^{1/3}\right)^{2} - R. \qquad (5)$$

It is not possible to obtain agreement with relation (3) by means of the procedure described above of taking $\emptyset^{*}$ and n_{ws} somewhat different from their experimental starting values.

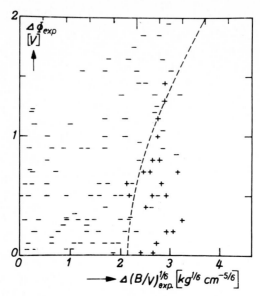

<u>Fig. 9</u>. Phase diagram analysis for binary systems of one of 27
transition metals and one of the metals Al, Ga, In, Tl, Sn,
Pb, Sb or Bi. The parameters are the experimental values
for $\emptyset$ and $n_{ws}^{1/3}$. (-) intermetallic compounds; (+) no compounds.

In fact in the course of these investigations the important step
forward has been the observation that, in treating alloys of d-
metals with p-metals as a special case, a nearly perfect description in terms of relation (3) was possible for the remaining
transition metal alloys.

In addition one should remember that we have already used
all our degrees of freedom, as far as the choice of $\emptyset^{*}$ and n_{ws}
is concerned, in figs. 4 and 7. All d-metals and p-metals have
got best fit $\emptyset^{*}$ and n_{ws} values already. Fig. 10 presents the
analysis of the sign of ΔH for this group of alloys in terms of
the recommended values for $\emptyset^{*}$ and $n_{ws}^{1/3}$. The main difference with
Fig. 9 is the much larger number of data points.

The hyperbola drawn in fig. 10 offers a value for Q_{o}/P as
well as a value for R/P. The first one, Q_{o}/P, has the same value
as in all other figures considered above. This result suggests
that the contribution represented by R is indeed an additional
energy effect that does not strongly interfere with the two
earlier derived contributions. We suggest that this additional

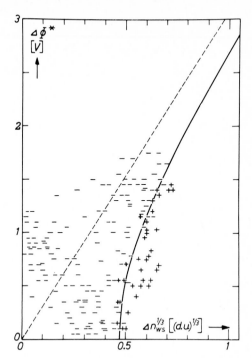

<u>Fig. 10</u>. As fig. 9, but now the parameters $\Delta\phi^{*}$ and $\Delta n_{ws}^{1/3}$ are
taken from the compilation of recommended values in
table 1.

negative energy effect occurring when d-metals have atoms of p-
metals as neighbours, is due to hybridization. One may imagine
that the correlation and exchange energies of mixed d-p electronic wave functions are more favourable than that of the original
d- and p-electrons. An argument in favour of a type of hybridization effect is found in the fact that also in liquid alloys
of d-metals with p-metals the extra negative term R is required,
although its value is about 25 percent smaller than for the
solid compounds. This is shown in fig. 11 which is the counterpart
of fig. 10 but now for liquid alloys. The number of data points
is limited but the correlation is quite convincing.

It is not surprising that in more detail the apparent
constant R/P varies with the valency of the polyvalent non-
transition metal. Fig. 12 is a more detailed version of fig. 10,
now including divalent metals too. It can be seen that the value

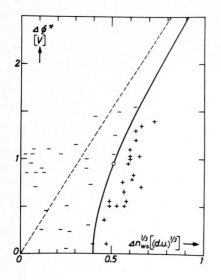

Fig. 11. Analysis of the sign of
ΔH for binary liquid
alloys of a transition
metal and a polyvalent
non-transition metal.
For details see ref. [10].

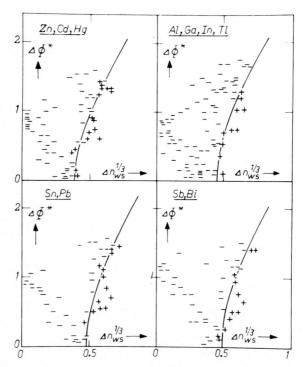

Fig. 12. Phase diagram analysis for binary alloys of a transition
metal and a polyvalent non-transition metal. Non-transition
metals of different valence are analysed separately which
results, in terms of relation (5), in somewhat different
values for the negative energy contribution R. For all
hyperbola Q_o/P is taken the same as before.

of R/Q_o increases gradually with increasing valency: the hyperbola drawn cuts the horizontal axis at a point determined by $(R/Q_o)^{\frac{1}{2}}$. The resulting values for R have been collected in table 2.

Finally we mention that also the sign analysis of ΔH for alloys of Cu, Ag and Au with p-metals agrees with a description of R being a hybridization effect. Also the three metals Cu, Ag and Au are to a certain extent still transition metals, as d-electrons contribute to the cohesive energy in the pure metals. Hence there will be some density of d-electrons at the boundary of the atomic cell. One therefore expects alloys of Cu, Ag or Au and p-metals to be described by relation (5) rather than (3) but with a relatively small value of R compared to the ordinary transition metals. This expectation is verified in fig. 13.

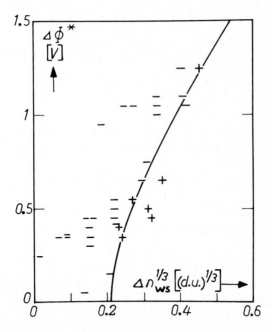

Fig. 13. Phase diagram analysis for binary alloys of Cu, Ag or Au and one of the metals Zn, Cd, Hg, Al, Ga, In, Tl, Sn, Pb, Sb or Bi. The number of data points is insufficient to permit a more detailed analysis, like that of fig. 12.

TABLE 2

The empirical constants P, Q_o and R in relation (5) for
the heat of formation of binary solid alloys.

Metal A	Metal B is a transition metal		
	R/P (eV^2)	Q_o/P $(eV^2/d.u.)^{2/3}$	P (V^{-1})
Transition metal	0	9.4	1.05
Alkaline metal or Cu, Ag, Au, Ca, Sr, Ba	0	9.4	0.85
Be, Mg	0.4	9.4	0.85
Zn, Cd, Hg	1.4	9.4	0.85
B, Al, Ga, In, Tl	1.9	9.4	0.85
C, Si, Ge, Sn, Pb	2.1	9.4	0.85
N, As, Sb, Bi	2.3	9.4	0.85

For solid alloys containing Cu, Ag, Au as metal B the values
of R are multiplied by 0.2. The difference with liquid alloys
is a factor 0.73 for R.
For alloys of two non-transition metals R=0.

V. SIMPLE RULES FOR ALLOYING

In this section we shall formulate the results of sections
III and IV in terms of simple rules for alloying of two metallic
elements. We go back to fig. 7 in which we analysed the 648 binary
systems in which metal A is a transition metal and metal B is a
transition metal or one of the metals Li, Na, K, Rb, Cs, Cu, Ag,

Au, Ca, Sr, Ba. The nearly perfect agreement obtained can be formulated as a rule for alloying of these metals.

A. If holds $|\Delta\phi^*| > 3.07 \times |\Delta n_{ws}^{1/3}|$ then the two metals alloy, i.e. either there are stable intermetallic compounds or at least one of the solid solubilities is larger than 10 at%. The units of ϕ^* and $n_{ws}^{1/3}$ are those of table 1. There are 304 binary systems among the above 648 to which this rule applies. There are two exceptions, Cr-U and Cu-Ta, so that the accuracy of rule A is 99%. The complementary situation is met in rule B:

B. If in a binary system there are no compounds, while in addition the mutual solid solubilities are smaller than 10 percent, then $|\Delta\phi^*| < 3.07 \times |\Delta n_{ws}^{1/3}|$. There are 112 systems to which this rule applies and there are two exceptions (again Cr-U and Cu-Ta). These are the only two that fall both in the category of rule A and that of rule B so that in fact we have two exceptions for 416 binary system. It is possible to reverse the statements as follows:

C. If in a binary system there are stable intermetallic compounds then $|\Delta\phi^*| > 3.07 \ |\Delta n_{ws}^{1/3}|$. Number of cases 268; four exceptions Cr-Mn, Fe-Mn, Pu-W and Th-Y.

D. If in a binary system $|\Delta\phi^*| < 3.07 \times |\Delta n_{ws}^{1/3}|$, there will be no stable intermetallic compounds. Here the number of cases is 177 with the same four exceptions as to rule C.

It is illustrative to visualize the above four rules by means of a graphical representation. In fig. 14 we make a map of the metallic elements in which each element has a position determined by its ϕ^* and $n_{ws}^{1/3}$ value. Now the two straight lines $\Delta\phi^* = \pm 3.07 \ \Delta n_{ws}^{1/3}$ are lines crossing at the position of the element choosen to be the reference metal. We get four sections, north, south, east and west. All metals in the north and south sectors will alloy readily with the reference metal, those in the east and west sectors will not. As an example we "predict" in fig. 14 the alloying behaviour of Fe. There are a large number of metals in both the north and south sectors with which Fe forms stable alloys indeed; there are no partners in the east sector but there are a lot of metals in the west sector. As mentioned above Fe-Mn forms an exception in the sense that there are compounds which are predicted to be absent.

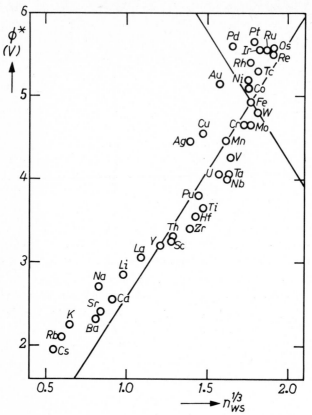

Fig. 14. Graphical representation of our simple rules for alloying. The lines drawn separate metals that have a negative heat of alloying with Fe from the metals that form alloys with Fe with a positive ΔH.

One notices immediately that this exception is not an important one; Mn lies on the line ΔH=0 in fig. 14. Note that in this figure it is explained why Fe forms compounds with Y (and the rare earth elements with an atomic volume near to that of Y) but not with La.

It will be clear that the graphical representation of fig. 14 does not apply to alloys of transition metals and p-metals. In that case one should make use of the two hyperbola's that correspond to ΔH=0 according to relation (5). In fig. 15 Pb is choosen as the reference p-metal. The two hyperbola calculated from an average value of R as determined in the foregoing section (fig. 10) are supposed to be the demarcation lines that separate the transition metals that form compounds with Pb from those transition metals that do not.

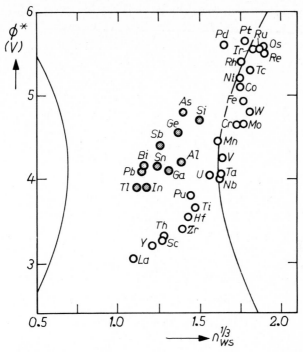

<u>Fig. 15.</u> Graphical representation of the alloying behaviour of
d-metals and p-metals (dashed ones). The hyperbola
separates d-metals that form stable intermetallic compounds with Pb from the transition metals that do not.

We may recall that the basic difference between alloys of
two transition metals and alloys of d-metals with p-metals or
alloys of two non-transition metals is that in the latter two
cases there is, or may be, an additional negative contribution
to the heat of alloy formation (hybridization energy of
Brillouin zone effect, respectively).An extra negative energy
does not interfere with our simple rules A and B and one therefore can expect these rules to be completely general. We have
investigated the validity of rules A and B for the 1326 binary
combinations of 52 metallic elements i.e. 27 transition metals,
the alkali metals, Cu,Ag and Au; Ca, Ba and Sr; Be, Mg, Zn, Cd
and Hg; Al, Ga, In, Tl, Sn, Pb, As, Sb and Bi. Rule A applies
to 578 systems, rule B to 233 systems. There are only five
exceptions (Cr-U, Cu-Ta, Ag-As, Tl-As and Cd-Sn).

In principle it is possible to formulate similar rules
for the occurrence of liquid immiscibility. The difference with
solid alloys is that for liquids T Δ S can be quite large.

The stability is so determined by $\Delta G < 0$ rather than by $\Delta H < 0$.
In order to have liquid immiscibility, ΔH at the equi-atomic
concentration must be larger than aT_m, where T_m is the higher
of the two pure metal melting points and a is a constant of the
order of 1 cal/g-at. A precise formulation of rules for
immiscibility of liquids requires a knowledge of the absolute
values of ΔH, the sign alone is not sufficient. However,
one can look upon the entropy term as an extra negative contribution to ΔG that favours miscibility. For this reason
rules A and B do apply to liquids:
A: If for two metals $|\Delta\emptyset^*|$ is larger than 3.07 x $|\Delta n_{ws}^{1/3}|$ the
two metals will be completely miscible in the liquid state.
B: If two metals are not completely miscible in the liquid
state than $\Delta\emptyset^* < 3.07$ x $|\Delta n_{ws}^{1/3}|$. Both rules apply to about
100 alloy systems for which there is experimental information.
There are two exceptions known, Ba-Ni and Li-Ni.

VI. ABSOLUTE VALUES

In sections III and IV we derived empirical values for
Q_0/P and R/P for different groups of alloys. In order to be able
to predict absolute values we have to determine P while furthermore the concentration dependence of ΔH should be known. We
start with the latter. Since the energy effect has been introduced as arising at the contact surface between dissimilar
atoms, its concentration dependence will be that of this surface
area. We introduce the surface concentration c_A^S of metal A

$$c_A^S = c_A V_A^{2/3} / (c_A V_A^{2/3} + c_B V_B^{2/3}) \qquad (6)$$

where c_A and V_A are the atomic concentration and molar volume
of metal A, respectively. As we showed previously [4] a general
expression for ΔH is

$$\Delta H/N_0 = f(c_A^S) \, g(c) \, [-Pe \, \Delta\emptyset^*)^2 + Q(\Delta n_{ws}^{1/3})^2 - R] , \qquad (7)$$

where N_0 is Avogrado's number and $g(c)$ stands for

$$g(c) = (c_A V_A^{2/3} + c_B V_B^{2/3}) / \bar{V}_m^{2/3}. \qquad (8)$$

For regular liquid or solid solutions $f(c_A^S)$ is

$$f(c_A^S) = c_A^S(1-c_A^S), \tag{9}$$

while for a series of ordered alloys it has been found empirically that $f(c_A^S)$ can be described by

$$f(c_A^S) = c_A^S(1-c_A^S) \left[1+8\left\{c_A^S(1-c_A^S)\right\}^2\right]. \tag{10}$$

The problem of predicting absolute ΔH-values is a matter of finding P. We reproduce here as fig. 16 a result for ΔH of

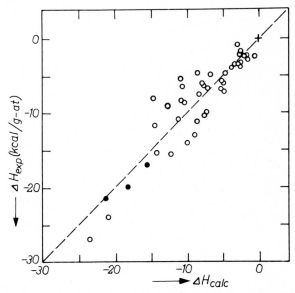

Fig. 16. Comparison of experimental and calculated values for the heat of formation of alloys of two transition metals. Calculation by means of relation (7) with R=0 and P=1.05 V^{-1}. For references to experimental data see ref. [4].

intermetallic compounds of two transition metals obtained in references [4] and [13]. In principle all experimental data available have been included except some U-compounds and some systems for which $|\Delta\phi^*/\Delta n_{ws}^{1/3}|$ is not sufficiently different (say less than 10%) from the critical value 3.07. The straight line drawn represents $\Delta H_{exp} = \Delta H_{calc}$; in the calculation we used P = 1.05 V^{-1}, the values for ϕ^*, n_{ws} and $V_m^{2/3}$ are from table 1.

The scatter of the data points not necessarily implies that
the calculation is incorrect; experimental uncertainties may
play their role too. In any case one gets an impression of the
accuracy that can be obtained in predictions of the heats of
formation of solid alloys of transition metals or those of
liquid alloys within the present model. Absolute values are
discussed in more detail elsewhere [13].

In the figs. 14 and 15 we developed a method to determine
immediately the sign of ΔH for alloys based on a given reference metal (Fe in fig. 14, Pb in fig. 15). This method can be
used to estimate absolute values too. If one draws fig. 17 on
a transparent and uses it in combination with figs. 14 and 15
one can read directly a value for ΔH of an imaginary ordered
alloy not far from the equiatomic composition (in fact
$c_A V_A^{2/3} = c_B V_B^{2/3}$). Fig. 17 gives absolute values for the case
of fig. 14 but it can be used for fig. 15 if one decreases all
ΔH-values indicated by -15 kcal/g-at. One can see this
already from the fact that the hyperbola that represents $\Delta H=0$
in fig. 15 carries the number +15 in fig. 17. The predictions
by means of fig. 17 are somewhat crude in the sense that
average values have been used for P and R as well as an average
value for $g(c)$. A more direct example of the predictive power
of the present model is given in table 3   where we compare
experimental and calculated values for ΔH of intermetallic
compounds of transition metals with Al.

We like to note here that, in addition to the compounds
of transition metals with all the other p-metals considered
already above, we do also predict ΔH-values for the transition
metal silicides, germanides, carbides and nitrides. They have
not been mentioned earlier in this paper, because we need to
include one additional positive energy contribution in order to
account for the fact that elementary Si, Ge, C and N are not
ordinary metals. When alloying Si, we first have to take Si
from its favourable, semiconducting, diamond structure and
transform it into an ordinary metal. This additonal positive
energy can be estimated from the unusually high heats of
melting of these elements to be about 8, 6 and 30 kcal/g-at,
per atom Si, Ge or C, respectively. Similarly in order to
describe ΔH for transition metal nitrides by means of relation
(7) and with the values of P, Q and R equal to those for the

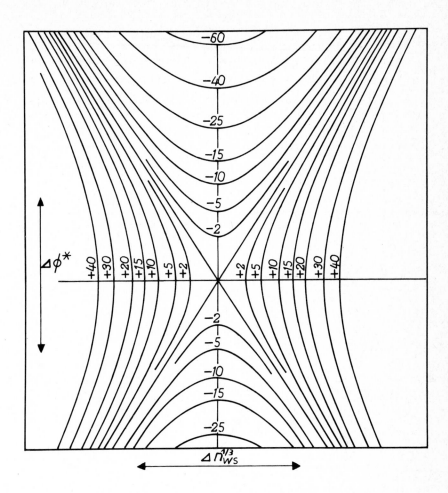

Fig. 17. The absolute values of ΔH (in kcal/g-at) for an ordered
solid alloy near the equi-atomic composition as a
function of Δø* and Δn$_{ws}^{1/3}$. The curves can be used in
combination with fig. 14 to estimate the heat of
formation of alloys that contain the reference element
at the origin. The curves can also be used in combination
with fig. 15 (d-metals and p-metals). Here all ΔH-values
indicated have to be decreased by 15 kcal/g-at.

TABLE 3

Comparison of experimental and calculated values for the heat
of formation ΔH of transition metal aluminides [13]. Numbers
in brackets are calculated, ΔH in kcal/g-at.

$ScAl_2$	$TiAl$	$V_{60}Al_{40}$	$Cr_{40}Al_{60}$	$MnAl$	$FeAl$	$CoAl$	$NiAl$
	- 8.9	-4;-9	-3.6	- 5.2	- 6.0	-12.9	-14.5
(-17.3)	(-15.6)	(- 9.5)	(-6.5)	(-10.8)	(- 7.7)	(-10.6)	(-11.9)
YAl	$ZrAl_2$	Nb_2Al	Mo_3Al	$TcAl_2$	$RuAl$	$RhAl$	$PdAl$
-21	-13.6*						-22.1
(-17.4)	(-16.3)	(-13.9)	(-3.5)	(- 8.5)	(-11.3)	(-14.9)	(-25.3)
$LaAl_2$	$HfAl$	Ta_2Al	WAl_4	$ReAl$	$OsAl$	$IrAl$	$PtAl$
-12.0							-24
(-15.7)	(-16.6)	(- 8.1)	(-2.2)	(- 8.4)	(-10.0)	(-13.9)	(-18.8)

$ThAl_3$	UAl_2	$PuAl_2$
- 8.7*	- 7.4	-11.3
(-14.3)	(-10.6)	(-12.7)

*   experimental value of ΔG.

other pentavalent metals As, Sb and Bi, we have to introduce a
large positive contribution to the enthalpy that takes care of the
transformation of molecular N_2 into imaginary metallic N. For
details see ref. [13]. Here is sufficient to recall that the
atomic model also accounts for transition metal compounds with
B, C, Si, Ge and N.

VII. COMPARISON WITH OTHER TREATMENTS

There is an interesting similarity between the present description of a positive contribution to ΔH of alloys in terms
of a discontinuity in the density of electrons at the boundary
between dissimilar atoms, and a much older description of interactions between atoms in a molecule as well as that of the heat
of mixing of non-plar liquids. One generally estimates the
interaction constant between two different atoms A and B in
a molecule AB to be equal to the geometrical average of the
interaction constants in the molecules AA and BB. This is
Berthelot's relation

$$e_{AB} = e_{AA}^{\frac{1}{2}} e_{BB}^{\frac{1}{2}} \tag{11}$$

where e_{AA}, e_{BB} and e_{AB} are the absolute values of the inter-action coefficients.

In terms of the heat of formation of the molecule AB relative to a mixture of AA and BB we find the heat of formation to be always positive (if we consider only stable molecules) since a geometrical average is smaller than a linear average:

$$- \Delta H/N_o = e_{AA}^{\frac{1}{2}} e_{BB}^{\frac{1}{2}} - (e_{AA} + e_{BB})/2 = -(e_{AA}^{\frac{1}{2}} - e_{BB}^{\frac{1}{2}})^2/2. \qquad (12)$$

Hildebrand and Scott [14], starting from relation (11), derived an expression for the positive heat of mixing of non-polar liquids that agreed quite well with experiment. For a liquid it is straight-forward to identify the interaction energies between atoms with the binding energy per unit volume or per unit atomic surface. Hildebrand and Scott write

$$\Delta H_{pos} = f(c_A^S) g(c) \bar{V}_m^{2/3} \left[(E_A^V/V_A^{2/3})^{\frac{1}{2}} - (E_B^V/V_B^{2/3})^{\frac{1}{2}} \right]^2. \qquad (13)$$

Here $f(c_A^S)$ and $g(c)$ have been defined in the foregoing section, V_m is the average molar volume and E_A^V and E_B^V are the heats of vaporization of the two liquids. Relation (13) can be equivalent to our expression for the positive term

$$(\Delta H/N_o)_{pos} = f(c_A^S) g(c) Q_o (\Delta n_{ws}^{1/3})^2, \qquad (14)$$

provided that the heat of vaporization per unit surface, $E^V/V_m^{2/3}$, is related to our density parameter n_{ws}. This relation is investigated in fig. 18; the line drawn corresponds to the linear dependence $E^V/V_m^{2/3} \sim n_{ws}$. Since there also is an approximate relation between n_{ws} and V_m (fig. 19, $V_m \simeq n_{ws}^{-0.6}$) we can rewrite relation (13) as

$$\Delta H_{pos} \sim n_{ws}^{-1.4} (\Delta n_{ws})^2 \simeq (\Delta n_{ws}^{1/3})^2. \qquad (15)$$

Comparison of absolute values from [13] or [14] using our experimental value for Q_o, learns that they agree within a factor 2 for all practical alloy cases.

Pauling [2] has suggested an expression for the heat of formation of ionic compounds

$$\Delta H = - 23M(X_A - X_B)^2 \quad \text{(in kcal/g-at)} \qquad (16)$$

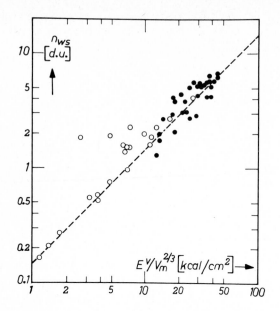

Fig. 18. The approximate linear relationship between the electron
density parameter of the present paper and the heat
of vaporization per unit atomic surface. All metallic
elements are included; filled circles represent transition
metals.

Here X_A and X_B are the electronegativities of two elements A
and B while M represents the number of shared electron pairs.
The electronegativities for metallic elements recommended by
Pauling, or different ones due to other authors, are on the whole
not very different from our $\emptyset^*$ values since as a matter of fact
any scale factor between $\Delta\emptyset^*$ and ΔX is acceptable. However,
the differences are significant.

The main problem that arises when one likes to apply
relation (16) for metallic compounds is the coefficient M. What
is the number of resonating bonds in a metal?

We remark that a result of the present work is that the
dominant contributions to the heat of formation do not depend on
crystal structure (e.g. for liquid alloys and solid alloys of
transition metals). Consequently the number of bonds is not likely
to be a useful concept for metallic alloys.

Like Pauling's description, also that by Phillips **and** Van
Vechten [15] is more suitable for semi-conducting compounds
then for metals. Phillips and Van Vechten have given a sophisti-

cated treatment of the ionic and covalency effects (read Brillouin-zone effects) for the $A^N B^{8-N}$ compounds. However, like was the case for the theory of Pauling, also within Phillips' description it is difficult to see how heats of formation of an alloy can be positive. In the literature this difficulty is usually overlooked. We want to emphasize that even if one considers heats of formation of compounds within a given crystal structure only (for instance NaCl or Laves phases) the theory should account for positive values for those combinations of metals that do not form compounds at all. If there are no compounds in a binary system the heat of formation of an imaginary compound necessarily must be positive [16].

In fig. 16 we have included a large number of ΔH-values taken from a recent paper by Brewer and Wengert [17]. It is note-worthy that the occurrence of compounds of unusually high stability among binary alloys of a transition metal from the left hand side of the series of transition metals and one of the metals with a nearly complete d-shell is accounted for by the present model in a simple way. The occurrence of these high stability alloys has been considered previously as a strong argument in favour of the applicability of the Engel theory for alloys. From relation (3) and the $\emptyset^{*}$ scale of table 1 one may learn that in the present model the stability is fully due to ionicity, while the positive term remains relatively small. In this process of charge transfer the electropositive metals Sc, Y, R.E., Ti, Zr or Hf loose electrons while their partner metals Ni, Rh, Pd, Ir or Pt complete their d-shells. This "ionic" picture can by no means be reconciled with the Engel-Brewer theory in which one assumes in fact charge transfer in the opposite direction. We conclude: either the Engel-Brewer description cannot be applied to alloys or the results of the present paper are just accidental.

A great number of experimental data on the heat of mixing of liquid alloys have been taken from the work of Predel's group (see ref. [10] for details). Predel already some time ago noticed the occurrence of a positive contribution to the heat of mixing of liquid alloys. He describes it as a size-mismatch term [18] and was able to demonstrate correlations between a positive heat of mixing and the difference in atomic volume of the two metals involved. We have shown here, for in-

stance, by demonstrating the existence of similarities between
ordered compounds and liquids, that the positive term is not
due to a difference in atomic size but to a difference in
electron density. Since fig. 19 shows that densities n_{ws} and
atomic volumes are correlated, it obviously is difficult to
distinguish the two points of view considering heats of mixing
of liquid alloys alone.

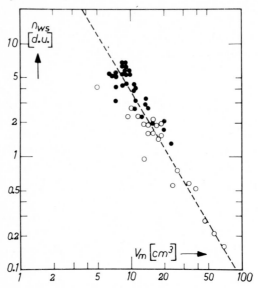

Fig. 19. The approximate relationship between molar volumes of
         metallic elements, V_m, and the density of electrons at
         the boundary of the atomic cell, n_{ws}. Open and filled
         circles represent non-transition metals and transition
         metals, respectively.

VIII. CONCLUSIONS

We have demonstrated that a model, in which energy effects
for metallic alloys are described in terms of interactions
between individual atoms, nicely accounts for the heat of
formation of solid as well as liquid alloys. It is assumed
that nearest-neighbour effects dominate; the contact surface
between dissimilar atomic cells is an important quantity.

The model, here introduced for binary alloys, can in
principle be applied to ternary alloys as well. As a first extension in this direction we have succesfully applied [19] the
atomic model to account for the enthalpy of formation of ternary

hydrides, like those of the family $LaNi_5H_6$.

The atomic point of view, taken in this paper to account for energies, can also be useful in relation to the density of states curves (number of electron states per unit energy) for alloys of two transition metals. We have recently shown [20] that, if one assumes that two metals in an alloy contribute individually to the density of states, a fair description can be obtained for the density of states at the Fermi energy and the transition temperature for superconductivity in solid solutions of transition metals.

ACKNOWLEDGEMENT

We wish to thank our colleagues, both at the Philips Research Laboratories, and at the University of Amsterdam, for many stimulating discussions on the present subject. The work performed at Amsterdam has been sponsored by the "Stichting voor Fundamenteel Onderzoek der Materie (F.O.M.)" and the "Metaal-instituut (T.N.O.)" at Delft.

References

[1] A.E. van Arkel, Moleculen en Kristallen (Van Stockum, 's-Gravenhage, 1961).

[2] L. Pauling, The Nature of the Chemical Bond (Cornell Univ. Press, Ithaca, N.Y., 1960).

[3] A.R. Miedema, F.R. de Boer and P. de Chatel, J. Phys. F (Metal Phys.) 3 (1973) 1558; A.R. Miedema, J. Less-Common Metals 32 (1973) 117.

[4] A.R. Miedema, R. Boom and F.R. de Boer, J. Less-Common Metals, to be published (1975).

[5] See, for instance, H. Brooks, Trans AIME 227 (1963) 546.

[6] K.A. Gschneidner, in: Solid State Physics, Vol. 16,(Academic Press, New York, 1964) p. 275.

[7] H.B. Michaelson, J. Appl. Phys. 21 (1950) 536.

[8] V.S. Fomenko, in: Handbook of Thermoionic Properties Ed. G.V. Samsonov (Plenum, New York, 1966).

[9] R. Hultgren, P.D. Desai, D.T. Hawkins, M. Gleiser and K.K. Kelley, Selected Values of Thermodynamic Properties of Binary Alloys (Am. Soc. Metals, Cleveland, Ohio, 1973).

[10] R. Boom, F.R. de Boer and A.R. Miedema, submitted to Acta Met.

[11] D.E. Eastman, Phys. Rev. B 2 (1970) 1.

[12] M. Hansen and K. Anderko, Constitution of Binary Alloys (McGraw-Hill, New York, 1958); and Supplements by R.P. Elliot (1965) and F.A. Shunk (1969).

[13] A.R. Miedema, J. Less-Common Metals, to be published.

[14] J.H. Hildebrand and R.L. Scott, The Solubility of Nonelectrolytes (Reinhold, New York, 1950) p. 133.

[15] See, for instance, J.C. Phillips, Rev. Mod. Phys. 42 (1970) 317.

[16] J.C. Phillips, J. Phys. Chem. Solids 34 (1973) 1051.

[17] L. Brewer and P.R. Wengert, Met. Trans. 4 (1973) 83.

[18] See,for instance,B. Predel and H. Sandig, Z. Metallk. 60 (1969) 208.

[19] H.H. van Mal, K.H.J. Buschow and A.R. Miedema, J. Less-Common Metals 35 (1974) 65.

[20] A.R. Miedema, J. Phys. F (Metal Phys.) 3 (1973) 1803; 4 (1974) 120.

Crystal Structure and Chemical Bonding in Inorganic Chemistry
Eds. C.J.M. Rooymans and A. Rabenau
© 1975, North-Holland Publishing Company, The Netherlands

THE IONIC MODEL AND THE ORIGIN OF LIGAND FIELDS

W.C. Nieuwpoort

Laboratory of Structural Chemistry, University of Groningen

Zernikelaan, Groningen, The Netherlands

I. INTRODUCTION

In this survey, two main aspects of the ionic model will be briefly discussed viz. the suggested charge distribution in compounds and the means it provides to calculate the cohesion energies of ion arrangements. Special mention of the Born respulsion and its origin will be made. Then we shall survey various approaches to the crystalfield problem arising when the ionic model is applied to ions with unfilled d and f shells. On the basis of recent self-consistent molecular orbital calculations we shall be led back to the ionic model and in particular to its Born repulsion part. Finally something will be said about the concept of covalency and the possible inadequacy of molecular orbital theory to take this fully into account. This is relevant to the interpretation of the observed reduction of free ion Racal parameters when an ion becomes bound in a complex, the "nephelauxetic" effect.

II. THE IONIC MODEL

The basic entities of this model [1,2] are charged spheres with more or less well defined dimensions and of a certain degree of hardness. We call these entities ions, however, we should be aware of the fact that these are conceptual quantities which need not have much to do with observable free ions. This can be seen by comparison with what is actually known about the charge distribution of stable ions, but most convincingly by realising that no doubly or higher charged anions seem to exist in nature,

while they can be employed without much difficulty in solids.
This is a clear warning against forming too literal pictures
of the charge distributions in solids as a superposition of ionic
distributions. There are other arguments against such simpli-
fications [3] and we shall see that theory indeed requires
substantial rearrangements of ionic distributions when the
equilibrium distances are approached.

Although the charge distribution picture provided by the
ionic model is not of too much value then, we all know its
tremendous power in calculating cohesion energies. The soundness
of such calculations from a quantum mechanical point of view
has been discussed some years ago by Blasse and the present
author [4,5]. This discussion was based on the quantum mechanical
virial theorem, which had been used earlier by Fröman and Löwdin
[6] to argue that the ionic model is fundamentally wrong
because it only deals with changes in potential energy and fails
to account for the changes in the kinetic energy of electrons
as is required by the theorem mentioned. In fact this argument
is incorrect because the repulsive "potential", which forms
an essential part of the ionic model, must be interpreted as
a pseudo-potential describing just the necessary electronic
kinetic energy change. This is easily seen by substituting
the ionic model energy expression

$$E = E_{rep} + E_{mad} \qquad (1)$$

into the virial relations

$$\Delta <T> = -E - R \frac{dE}{dR} \qquad (2a)$$

$$\Delta <V> = 2E + R \frac{dE}{dR} \qquad (2b)$$

where $<T>$ and $<V>$ represent respectively the average electronic
kinetic energy and the average potential energy including the
nuclear repulsion energy.

It is amusing to observe this consistency of the "classical"
ionic model with quantum mechanical requirements in contrast to
the elementary treatments of the covalent bond in H_2 or H_2^+, be
it the Heitler-London or LCAO-MO method, which do not obey the
virial theorem unless a scale-factor is introduced in the wavefunction the physical interpretation of which is not easy, however [7].

The kinetic energy change described by the short range Born-repulsion term implies of course a change in the electronic wavefunctions of the constituent ions. A major contribution to this change arises from the Pauli-principle, the effects of which become noticeable when the ion wavefunctions start to penetrate each other. In the usual orbital approximation to these wavefunctions the effects are accounted for by mutual orthogonalization of the orbitals of the respective ions. In ref. [4] and [5] the changes in the cohesion energy as a function of distance are discussed in more detail and it is made plausible why in many cases the cohesion energies can be safely estimated from the Madelung contribution alone. The examples given there show that for this purpose it is not at all necessary to deal with compounds that are "ionic" in the usual sense.

III. CRYSTAL FIELD THEORY

It is well known that when the ionic model is applied to systems containing cations with open d- or f-shells, these shells break up because of the non-spherical symmetry of their surroundings. A basic theory to deal with this problem, the crystal field theory, has been formulated many years ago by Bethe [8]. It has been applied first to the magnetic behaviour of Rare Earth [9] and transition metal [10] compounds and later also to optical properties. Further developments [13-17] have led to the broad area of applications as we know it today. Bethe had emphasized the group theoretical analysis but in the application of the theory it became customary to approximate the surroundings of the metal cations by point charges or point dipoles so that the perturbing fields could be explicitly defined in electrostatic terms. In the course of time it appeared, however, that a number of observations such as the spectrochemical series, the transferred hyperfine effect, and the reduction of free ion Racah parameters could not be rationalized without the introduction of electron delocalization in the theory.

In most applications this is accomplished by replacing the electrostatic model by a semi-empirical molecular orbital (m.o.) model, which is more flexible and allows at least a qualitative explanation of most, if not all, phenomena. It lacks, however, the quantitative straightforwardness of the simple electrostatic model. In Tables 1a and 1b the essential features of the various forms of semi-empirical theory are summarized.

TABLE 1a
Semi-empirical ligand field theories

1. first order perturbation theory

   or

   simple molecular orbital theory

2. only "d" electrons explicitly treated

3. $\hat{H} = \sum h(i) + \sum_{i>j} g(ij) + \sum_i V_{so}(i) + \sum_i V_L(i)$

   $= \sum_i h(i) + V_{central}(i) + \left| \sum_{i<j} g(ij) - V_{central} \right| + \cdots$

   $= \hat{H}_{central} + V_{ee} + V_{so} + V_L$

4. ψ : single configurational, determinantal type

TABLE 1b
Further classification ligand field theories

type	character	pert	manifold	parameters
weak field	first order pert. theory	V_L	$\lvert d^n, L, S \rangle$ $\lvert d^n, J \rangle$	Δ (10Dq) (A), B, C
strong field	first order pert. theory		$\lvert t_2^x(d) e^y(d), \Gamma, S \rangle$	ΔB, C
	one center m.o.	V_{ee}	$\lvert t_2^x(d) e^y(d'), \Gamma, S \rangle$	Δ
	many center m.o.		$\lvert t_2^x e^y, \Gamma, S$	$3(A, B, C)$ D

IV. QUANTITATIVE INTERPRETATIONS OF 10 Dq

The application of the theory in parametrized form was and is very succesful as long as one does not probe too deeply into the meaning of the parameters and hence into the origin of the ligand field. In particular the octahedral ligand field splitting parameter 10 Dq has been the subject of quantitative investigations. In recent years a number of attempts have been made up to obtain this quantity for the perovskite $KNiF_3$ by first principles or ab initio approaches [18-27]. The essentials of such treatments are listed in Table 2. In our own work [26,27]

TABLE 2

Ab-initio approaches 1. variational treatments 2. complete hamiltonian		
type	character	trial functions
weak field	Heitler-London	a. $A\lvert d^n, L, S >\rvert$ ligands> + b. charge transfer states
strong field	SCF-MO	a. $\lvert \ldots\ldots t_2^x\, e^y \ldots >$ + b. configuration interaction

in this direction a number of calculations were carried out to obtain 10 Dq for $KNiF_3$ specifically in terms of the electro-statically computed value and various corrections to it. Two of the conclusions may be recalled here.

First of all the mutual orthogonalization of metal 3d-orbitals and ligand orbitals contributes much more to the splitting than the electrostatic terms: point charges ~ 1400 cm^{-1}, orthogo-nalized ions ~ 4000 cm^{-1}, final SCF value ~ 5500 cm^{-1}. Secondly the possible improvement on the SCF level was estimated to be on the order of a few hundred cm^{-1}, so that the single con-figurational m.o. model would be inadequate to completely reproduce the experimental value of 7250 cm^{-1}. Similar con-clusions have been reached [28] for the more covalent tetrahedral complex $NiCl_4^{2-}$: point charges ~ 500 cm^{-1}, orthogonalized ions ~ 1200 cm^{-1}, final SCF value ~ 2300 cm^{-1}, experiment ~ 3600 cm^{-1}. The first conclusion means that the ionic model can still serve as a reasonable starting point to discuss the ligand field but that it is imperative to include the anisotropic part of the Born repulsion in one's consideration. Clearly, qualitative considerations on the basis of the spatial extension of metal d-orbitals remain the same as for the point charge model as long as one is dealing with negative charges. The second conclusion suggests that in spite of its qualitative success the M.O. model does not sufficiently describe the "covalency" i.e. all de-localization effects beyond those imposed by orthogonalization (or "overlap distortion") effects. For semi-quantitative purposes the Heitler-London + charge-transfer states approach used by Rimmer and Hubbard might well be preferable.

V. RACAH PARAMETERS

The possible inadequacy of the M.O. model may be of
particular relevance to the interpretation of the observed
reduction in magnitude of the electron repulsion parameters B and
C when a metal ion becomes bound in a complex. Especially
Jörgensen has attributed this effect to the change in open shell
orbital character from pure metal to a mixture of metal and
ligand functions. From our own and other ab initio calculations
it would appear that this is not a sufficient explanation. In
Table 3 our results for NiF_6^{4-} and $NiCl_4^{2-}$ are displayed. This
matter is more extensively discussed in ref. [26] and [27].

TABLE 3
Experimental and theoretical Racah parameters Ni(II)

	$B(cm^{-1})$		$C(cm^{-1})$	
	exp	theor	exp	theor
free ion	1030	1315	4850	4840
NiF_6^{4-}	955	1280	4234	4810
$NiCl_4^{2-}$	750	1308	3100	4290

References

[1] A.E. van Arkel and J.H. de Boer, "Chemische Binding als Electrostatisch Verschijnsel", (Centen, Amsterdam, 1930).

[2] A.E. van Arkel, "Moleculen en Kristallen" (Van Stockum, 's-Gravenhage, 1961).

[3] J.C. Slater, "Quantum Theory of Molecules and Solids" Vol. 2 (Mc Graw Hill, New York, 1965) p. 105.

[4] W.C. Nieuwpoort and G. Blasse, Chemisch Weekblad 63 (1967) 497 (in Dutch).

[5] W.C. Nieuwpoort and G. Blasse, J. Inorg. Nucl. Chem. 30 (1968) 1635.

[6] A. Fröman and P.O. Löwdin, J. Phys. Chem. Solids 23 (1962) 75.

[7] K. Ruedenberg, Rev. Mod. Phys. 34 (1962) 326.

[8] H. Bethe, Ann. Physik 3 (1929) 133.

[9] W.G. Penney and R. Schlapp, Phys. Rev. 41 (1932) 194, 666.

[10] J.H. Van Vleck, Phys. Rev. 41 (1932) 208.

[11] D. Polder, Physica 9 (1942) 709.

[12] E. Ilse and H. Hartmann, Z. Physik. Chem. (Leipzig) 197 (1951) 239.

[13] L.E. Orgel, J. Chem. Soc. (1952) 4756.

[14] Y. Tanabe and S. Sugano, J. Phys. Soc. Japan 9 (1954) 753.

[15] J.S. Griffith, "The Theory of Transition-Metal Ions", (Cambridge University Press, 1961).

[16] C.J. Ballhausen, "Introduction to Ligand Field Theory", (Mc Graw Hill, New York, 1962).

[17] C.K. Jörgensen, "Absorption Spectra and Chemical Bonding in Complexes" (Pergamon, New York, 1962).

[18] S. Sugano and R.G. Shulman, Phys. Rev. 130 (1963) 517.

[19] R.E. Watson and A.J. Freeman, Phys. Rev. 134 A (1964) 1526.

[20] P.O. Offenhartz, J. Chem. Phys. 47 (1967) 2951.

[21] J. Hubbard, D.E. Rimmer and F.R.A. Hopgood, Proc. Phys. Soc. (London) 88 (1966) 13.

[22] D.E. Ellis, A.J. Freeman and P. Ros, Phys. Rev. 176 (1968) 688.

[23] H.M. Gladney and A. Veillard, Phys. Rev. 180 (1969) 385.

[24] J.W. Richardson, D.M. Vaught, T.F. Soules and R.R. Powell, J. Chem. Phys. 50 (1969) 3633.

[25] J.W. Moskowitz, C. Hollister, C.J. Hornback and H. Basch, J. Chem. Phys. 53 (1970) 2570.

[26] A.J.H. Wachters, Thesis, Groningen 1971.

[27] A.J.H. Wachters and W.C. Nieuwpoort, Phys. Rev. B 5 (1972) 4291.

[28] T.J.M. Smit, unpublished results of this laboratory.

Crystal Structure and Chemical Bonding in Inorganic Chemistry
Eds. C.J.M. Rooymans and A. Rabenau
© 1975, North-Holland Publishing Company, The Netherlands

TRIPLE INTERACTION MODELS IN IONIC SOLIDS

L. JANSEN

Battelle Institute, Advanced Studies Center,

CH-1227 Carouge-Geneva, Switzerland

and

Institute of Theoretical Chemistry, University of Amsterdam,

The Netherlands.

ABSTRACT

After a review of the history of many-ion (and many-atom) forces in dense media, a simple model of "effective" electrons is developed to evaluate such forces, of exchange type, on the basis of exchange perturbation theory. The model is applied to the following problems: A. crystal stability of alkali halides; B. crystal stability of II-VI and III-V compounds, with closed electron shells on the ions; C. crystal stability of compounds AX_2, with closed electron shells on the ions; D. crystal stability of compounds AX, with closed 3d-shells on the cations; E. deviations from the Cauchy relations between elastic constants of ionic solids; F. magnetic ordering in non-conducting ionic solids with paramagnetic 3d- or 4f-shell cations. In all these cases, it is found that a Hund-Born-Mayer model, *supplemented by three-ion exchange forces*, gives good agreement with experiment. The necessity of considering also covalent (homopolar) binding in the solids is never encountered. Finally, two problems are discussed on the basis of many-*atom* exchange interactions: G. the stability of rare-gas halides, and H. the interpretation of rotational barriers in simple organic and inorganic molecules.

I. INTRODUCTION

The origin of the concept of many-ion forces in solids can be traced back to Kramers [1], who in 1934, on the basis of a simple model of three centers and four electrons, analyzed the indirect exchange (later called "superexchange") interaction between two paramagnetic (3d-) cations (Ni^{2+}, Mn^{2+}, etc.) in a non-conducting solid *via* a diamagnetic anion (e.g. O^{2-}). The first experimental indication for the occurrence of super-exchange was established, also in 1934, by Becquerel, de Haas and van den Handel [2] in measurements of magnetic properties

of the mineral tysonite (a fluoride of La and Ce, with Nd and
Pr as impurities). In 1946, de Klerk [3] obtained further evidence
for superexchange from measurements of the specific heat of
copper potassium sulfate at temperatures below 1°K. This specific
heat was found to be much larger than that due to *isolated* unpaired spins on the Cu^{2+}-cations, pointing to an exchange
coupling between different cations. As the Cu-ions are
relatively far apart, the coupling must be of the indirect-
exchange type (a three-ion exchange interaction). In 1949, Shull
and collaborators [4] published the first results of a series
of neutron-diffraction analyses on non-conducting 3d-solids at
low temperatures. The evidence of magnetic ordering of spins on
different paramagnetic cations was conclusive proof for the
existence of the superexchange phenomenon. Since then, magnetic
order and superexchange interactions have formed a major domain
of experimental and theoretical solid-state research. A considerable number of reviews on theories of superexchange have
appeared in the literature; we here refer to only a few of them
[5-7].

     Whereas the first discussions of many-ion forces were con-
cerned with open-shell cation-cation indirect exchange, later
attention was also drawn to the possible existence of *many-atom*
forces (between rare-gas atoms, primarily) and of interactions
between more than two *closed-shell* ions (in first instance, those
of the alkali halides in the solid state). The first analysis of
three-atom interactions was undertaken by Axilrod and Teller [8]
on the basis of third-order perturbation theory without exchange.
The resulting so-called "triple-dipole" forces between three
rare-gas atoms are a straightforward extension of the London-van
der Waals dipole-dipole forces in second order. They were later
applied by Axilrod [9] to the stability problem of solids of the
heavy rare-gas atoms: pair potentials between the atoms had
failed to explain the observed stability of the face-centered
cubic configuration for solid Ne, Ar, Kr and Xe, relative to that
of the hexagonal close-packed structure. Axilrod's attempt failed
in that the triple-dipole forces were found to be insufficiently
stereospecific to decisively favor the cubic configuration. Numerous other applications of these long-range three-atom interactions have in the meantime appeared in the literature, often
while ignoring the fact that they represent only the *far end*
"tail" of a three-atom potential [10,11].

Margenau [12] had already pointed out in 1939 that also exchange interactions (also called "first-order" and "overlap"
forces) between closed-shell atoms or ions are not pairwise
additive. The first explicit calculation was carried out by
Löwdin [13] for alkali-halide crystals, followed by Rosen [14]
and Shostak [15] for three helium atoms. For the He-system, the
three-atom interactions are too small to be of any importance in
applications. Löwdin's analysis of many-ion forces in solid
alkali halides, is, however, of great interest in connection
with the well-known problem concerning the observed stability
of the B2(CsCl)-structure for CsCl, CsBr and CsI at standard
temperature and pressure. According to the Hund-Born-Mayer (HBM)-
model, otherwise remarkably successful in correlating different
properties of ionic solids, the stable structure for all alkali-
halide solids should be the B1(NaCl)-configuration. Löwdin found
a considerable three-ion component of the interaction energy,
amounting to 10-20 kcal/mole, with negative sign, for alkali halides with small positive and large negative ions. However, these
three-ion forces decrease rapidly as the ions approach equal size,
contrary to what is required to explain B2-stability for the
heavy Cs-salts. Moreover, the three-ion crystal energy is practically proportional to the Madelung constant of the lattice. Since
these constants for the B1- and B2-configurations differ by only
1%, the structure sensitivity of the effect is much too low to
account for B2-stability (the HBM-model predicts an energy
difference of several kcal/mole in favor of B1). Löwdin's results
for deviations from the Cauchy relations between elastic constants
show better agreement with experiments. These deviations,
measured at very low temperatures, for crystal structure of cubic
symmetry, must be due to many-ion forces in the solids. They
constitute, therefore, *direct* experimental proof for the
occurrence of *some* type of many-ion interactions. We mention also
a semi-classical analysis by Dick and Overhauser [16], based on
electrostatic interactions involving "exchange charges" which
arise as a consequence of the Pauli-principle. For two mutually
not-overlapping pairs of ions this effect gives rise to four-ion
forces, and to three-ion forces between a pair of neighbors and
a more distant ion. Such interactions are, however, not
applicable to crystal stability.

The applications of many-atom or many-ion exchange forces,
in problems of stability of different spin patterns in solids

with paramagnetic cations and in problems of crystal stability,
are based on the expected high stereospecificity of such interactions. The consequences can already be demonstrated when we
consider the simplest possible case, that of a comparison between
the face-centered cubic (fcc) and the hexagonal close-packed (hcp)
structures for rare-gas solids. In both structures, a central
atom is surrounded by 12 nearest neighbors and 6 second neighbors;
the first difference between fcc and hcp occurs only in the third
shell. If the interactions were pairwise additive, then the
difference in crystal energy would be negligibly small (of the
order $10^{-2}\%$, i.e. ~ 1 cal/mole).

Let us suppose that three-atom exchange forces must also be
considered. Since the atomic separations are large compared to
distances between ions in their solids, and since exchange interactions are of short range, we may here limit ourselves to the
first shell of atoms around a central rare-gas atom. There are
12 x 11/2 = 66 combinations of a central atom and two atoms from
the first shell, in both structures. It appears that 57 of these
triangles are the same between fcc and hcp, but nine are different
(the fcc lattice is centro-symmetric, but the hcp is not). Thus,
differences in three-atom interactions must be expected already
from the *first* shell.

In the HBM-model of ionic solids, the ions interact through
their net charges (Madelung energy) and through repulsive exchange
forces between their closed electron shells; these forces are
assumed to be additive in pairs. Deviations from additivity must
become more pronounced if the ions are large and also if their
separations become small, i.e. if their net charges increase.
Characteristically, the agreement between predictions of the model
and experiments becomes poorer in the same direction. If we
consider II-VI and III-V solids with closed electron-shell ions
(BeO, CaS, MgTe; BN, AlAs, etc.), then it is found that the B2-
structure no longer occurs. In addition to B1, two new structures
appear, namely B3(sphalerite) and B4(wurzite); both have tetrahedral coordination. Since their stability cannot be reasonably
explained on the basis of the HBM-model, and since B3 is identical,
for identical atoms occupying the two fcc-sublattices, with the
diamond structure, one assumes that *covalent bonding* is responsible
for the occurrence of B3 and B4. The only quantitative significance of this concept for the description of ionic solids is that
it measures an observed deviation from results obtained on the

basis of the HBM-model, in much the same way as the concept of
"correlation energy" in quantum chemistry does so with respect
to the Hartree-Fock solution for a molecular problem. However, in
addition to the fact that the concept of covalency explains
everything and nothing, it often carries the implication of
appealing to "chemical intuition" (electron-pair bonding) and,
thus, of impeding more detailed analyses. We have attempted to
obtain more insight concerning the limits of validity of the
HBM-model, assuming that the concept of *many-ion exchange forces*
in the solids may generally lead to a quantitative explanation of
observed crystal stability. After a brief description of the model
used, we will report a number of results and discuss applicability
of the theory to different domains of chemistry and physics.

II. THE MODEL

Consider an assembly of ions (cations and anions) in a solid.
We assume that the crystal energy can be written as the sum of an
"electrostatic" and an "exchange" component and that, as in the
HBM-model, the electrostatic part is given, with sufficient accuracy, by the total Coulomb interaction between the net ionic
charges (Madelung energy). The exchange part of the interaction
energy is written as a cluster expansion: two-ion exchange, three-
ion forces, etc. It is further assumed that, for the solids to be
considered, this expansion may be terminated with the *three*-ion
term.

Let $\underline{a}$, $\underline{b}$, $\underline{c}$ denote a cluster of three ions in an arbitrary
geometric configuration. We consider $\underline{a}$ to be the central ion;
since exchange forces die out rapidly with increasing ion-sepa-
ration, $\underline{b}$ and $\underline{c}$ are ions close to $\underline{a}$. We denote by $E(abc)$ the
interaction energy for the system of three ions and by
$E^{(o)}(ab)$, $E^{(o)}(ac)$ and $E^{(o)}(bc)$ the interactions between $\underline{a}$ and $\underline{b}$
with $\underline{c}$ removed, between $\underline{a}$ and $\underline{c}$ with $\underline{b}$ removed, and between $\underline{b}$
and $\underline{c}$ with ion $\underline{a}$ taken away, respectively. For the evaluation
of these four quantities, we consider the ions to be *electrically
neutral*. The three-ion energy $\varepsilon(abc)$ is given by

$$\varepsilon(abc) = E(abc) - \left\{ E^{(o)}(ab) + E^{(o)}(ac) + E^{(o)}(bc) \right\}. \tag{1}$$

In the problem of evaluating exchange forces between two para-
magnetic cations (a and $\underline{c}$) via a diamagnetic anion ($\underline{b}$), we de-
fine an *effective* cation-cation interaction $\varepsilon(ac)$ as follows

$$\varepsilon(ac) = E(abc) - \left\{ E^{(o)}(ab) + E^{(o)}(bc) \right\} ; \qquad (2)$$

this quantity contains, from its definition, also direct cation-
cation exchange.

The interactions are evaluated on the basis of perturbation
theory, in first and second orders, with the isolated ions as the
unperturbed system. Since the forces are of exchange type, conventional Rayleigh-Schrödinger (RS) perturbation theory cannot be
used. The reason is that the RS-method starts from *simple product*
unperturbed wavefunctions, which are thus not adapted to the permutation symmetry of the Hamilton operator for the system. It
can, however, be shown [17,18] that RS-perturbation theory can
be reformulated in such a way as to render it applicable for the
evaluation of exchange forces; this involves a re-definition
("symmetrization") of the Hamilton operator H_o for the unperturbed system and of the perturbation H'. For details we refer
to the literature [17,18]. The expressions for first- and second-
order interactions are evaluated on the basis of an *effective -
electron* model similar to the one used by Kramers in his analysis
of superexchange [1]. We replace the closed shells of electrons
on each ion by one effective electron, and place the spins of
these electrons on the three ions $\underline{a}$, $\underline{b}$ and $\underline{c}$ parallel (otherwise
Heitler-London chemical bonding occurs). One can also take *two*
spin-paired effective electrons per ion, but this only complicates
the calculations and provides no new information. The (symmetri-
zed) perturbation H' contains the electrostatic interactions
between the three effective electrons, between each electron and
the nuclei of the two ions with which it is not associated (for
each possible "labeling" of the electrons), and between the
nuclei (charge plus one) themselves. If paramagnetic cations are
involved, then we adopt Kramers' three-center, four-electron
model, with one effective electron per cation and two spin-paired
effective electrons per anion in the unperturbed ground state.

For the unperturbed wavefunction $\phi_o(r)$ of each effective
electron, we choose a (normalized) simple-Gaussian form

$$\phi_o(r) = (\beta/\pi^{1/2})^{3/2} \exp(-\beta^2 r^2/2), \qquad (3)$$

where $\underline{r}$ is the distance between the effective electron and its
nucleus, and where β is a parameter characteristic for the "size"
of the ion (the parameter β for a paramagnetic cation charac-

terizes its "magnetic size"). For ionic solids of composition
C_mA_n, where the symbol C stands for cation, A for anion, we
need two such Gaussian parameters: β for the anion and β' for the
cation. This leads to a *relative-size* parameter $\gamma = (\beta'/\beta)^2$ (the
square is chosen for convenience); for $\gamma < 1$ the cation is larger
than the anion, whereas for $\gamma > 1$ the situation is reversed. Values
for the Gaussian parameters can be estimated on the basis of
diamagnetic susceptibilities for the ions, relative to those
values for the rare-gas atoms. For an atom (ion) with one
effective electron, the diamagnetic susceptibility is proportional
to the expectation value $<r^2>$ for the electron, i.e. proportional
to β^{-2}. If one determines a gauge-value of β for one rare-gas
atom, then the β-values for the other atoms, as well as for all
closed-shell ions, can be calculated. We have chosen the Ne-atom
as a reference and determined β such that it reproduces optimally
both the short-range (repulsive) *and* the long-range (attractive)
parts of the Ne-Ne potential [19]; this gauge-value is 1.1 $\overset{\circ}{A}^{-1}$.
The particular choice is based on the observation that solid neon
is the lightest rare-gas solid stable in the fcc-configuration.
Thus, the origin of the preference for the fcc-configuration of
rare-gas crystals is most likely to be found already in the
interaction between a Ne-atom and its neighbors. Detailed calculations [19] showed that, for interactions between two Ne-atoms,
only *single*-exchange effects must be taken into account. This
result was the basis for the effective-electron model as applied
to crystal stability. It should be noted that accurate β-values
are not essential in connection with crystal stability and magnetic ordering; we are here primarily interested in the *trend* of
the results as a function of β .

The parameter γ takes the place of the quantity $(r_-/r_+)^2$
in the Goldschmidt-model, in which the ions are represented by
rigid spheres. That an explicit size-parameter should be added
to the HBM-model is to be expected in view of the fact that the
Goldschmidt-model yields remarkably closer agreement with experiment regarding crystal stability than the theoretically much-
better founded HBM-description. The variables occurring in the
theory are then: 1) crystal structure, 2) βR_o, with β the Gaussian
parameter for the anion and R_o the nearest-neighbor cation-anion
distance, and 3) the relative-size parameter γ . First-order
perturbation energies are calculated in units $e^2\beta$, second-order

results are obtained in units $e^2 \beta(e^2 \beta/E_{Av})$, where E_{Av} is an
average (Unsöld) excitation energy for the configuration (abc) of
the three ions considered. For a typical value $\beta = 1$ Å^{-1},
$e^2 \beta/E_{Av} \approx 1$ for $E_{Av} \approx 15$ eV. We can assume that the highest value of
$e^2 \beta/E_{Av}$ is of the order 1. From a comparison with other methods,
Dalgarno and Lynn [20] have established for H_2^+ and H_2 that E_{Av}
increases with decreasing distance between the nuclei. More
recently, van der Avoird [21] carried out an exchange perturbation analysis for H_2^+; E_{Av} for the antibonding state was found
to increase by a factor of more than ten upon decreasing the
internuclear distance from seven to three atomic units. In extra-
polating this property of E_{Av} to ionic solids, we may expect that,
because of electrostatic compression, E_{Av} has in general a high
value, suppressing second-order interactions.

III. RESULTS

We have applied the above model of many-atom or many-ion exchange forces to the problem of stability of rare-gas solids [22],
alkali-halide crystals [23], solids of II-VI and III-V compounds [24], ionic solids of composition AX_2 (CaF_2, SiO_2, OCs_2,
etc.) [25], to polymorphism of solids of compounds with closed d-
shell cations (CuF, AgBr, ZnO, CdTe, etc.) [26], to the evaluation
of deviations from the Cauchy relations in ionic solids [26,27],
to magnetic ordering in non-conducting solids with paramagnetic
(3d- or 4f-) cations [28], to the problem of stability of rare-gas
halides [29], and to the interpretation of rotational barriers in
simple organic and inorganic molecules [30].

We will now discuss some of the results, with principal reference to ionic solids and to the information they provide
relating to   many-ion forces.

A. *Stability of Alkali-Halide Crystals*   [23]: The problem, we
recall, is to explain the observed B2-stability for solid CsCl,
CsBr and CsI at normal temperatures and pressures or, more
generally, to account for the too-high stability of the B1-struc-
ture as determined on the basis of the HBM-model. To calculate
three-ion exchange forces, we start from a central ion (anion or
cation) and consider three-ion configurations, with two ions from
neighboring shells. The electrons of each of the three ions are
represented by one effective electron, and the spins of these

three electrons are parallel. The calculations show that only
isosceles triangles of ions, with the central ion at the apex,
need be considered, i.e. with the two ions from the *same* shell of
neighbors. Further, it is found that here only neighbors from
the first and second shells contribute to the three-ion crystal
energy.

In the following Table I we list general properties of the
three-ion forces as they can be inferred from the calculations.
They refer to alkali halides where the cations are not very much
smaller than the anions (i.e. except for the Li-salts and NaI).

Table I

Comparison between three-ion exchange interactions in the structures B1(NaCl) and B2(CsCl).

---

(1) Three-ion interactions between a central ion and two ions
from the *first* shell increase the attractive forces in the
crystal. This increase is larger in the B2- than in the B1-
lattice, i.e. this type of exchange stabilizes the B2-confi-
guration;

(2) Three-ion interactions between a central ion and two ions
from the *second* shell increase the repulsive forces in the
crystal. This increase is smaller in B1 than in B2, so that
this type of exchange stabilizes the B1-lattice;

(3) The order of magnitude of the three-ion exchange is up to a
few percent of the lattice energy.

On the basis of these general rules we can understand the
observed stability of the B2-structure for CsCl, CsBr and CsI:
the B2-lattice can become more stable, according to (1), for
heavy ions of about the same size. Then, in addition, the distance between the central ion and the second shell is large, i.e.
effect (2) is likely to be relatively small. Quantitative calculations confirm these conclusions [23]. It is found that the
difference in favor of B2 for CsCl, CsBr and CsI is larger than
the HBM-barrier, in favor of B1, by about 1 kcal/mole. The γ-
values for these salts are 0.81 for CsCl, 1.11 for CsBr and
1.44 for CsI. The F^--ion is too small to render B2 stable for
CsF. The calculated pressures of transition from the B1- to the
B2-structure are found to be in much better agreement with expe-

rimental values than those obtained on the basis of the HBM-model
alone. It follows that the HBM-model, *supplemented* by three-ion
exchange forces, can indeed account for B2-stability of the
heavy Cs-salts.

B. *Stability of Solids of II-VI and III-V Compounds* [24]: Here,
the lowest γ-value for II-VI compounds is 1.18 (BaO), followed
by 1.22 (SrO); for III-V compounds the lowest value is 1.78 (LaN),
followed by 2.64 (YN). Thus, only BaO and SrO could in principle
be stable in the B2-configuration. However, due to strong electrostatic compression, several further shells of ions around a central
ion must be considered for the evaluation of the three-ion energy.
The first shell favors B2 over B1, as for alkali-halide stability.
The second shell is closer to the central ion than in the solid
alkali halides; it also favors the B2-configuration. However,
in the "crowded" B2-structure, three further shells of ions are
important; all these favor the B1-configuration. (The number of
shells to be considered in the B1-lattice is only three.) It is
found that the effect of these further shells in B2 *rules out*
stability of that structure for II-VI and III-V compounds.

There are, however, two further structures observed with a
number of II-VI and III-V compounds, namely, the B3(sphalerite)
and B4(wurtzite) configurations, both with coordination four. As
was remarked in the Introduction, the occurrence of B3 and B4 is
often considered a manifestation of covalent bonding between
cations and anions in the solid. It may be noted that no B3-B4
polymorphism is observed with these compounds. Neither on the
basis of covalent bonding nor in the Goldschmidt-model is it, of
course, possible to distinguish between stability of B3 and B4.

What does the three-ion exchange model have to say concerning
stability of these structures? First of all, B3 and B4 are observed only if the value of the parameter βR_o is small and that of
the relative-size parameter is large: $\beta R_o \leqslant 1.0$; $\gamma \geqslant 15$. The high
values of γ imply that three-ion exchange forces involving one or
more cations are very small and can be neglected relative to the
three-anion forces. B3-stability is quite generally associated
with the lowest values of βR_o and the highest values of γ . Because of the low values of βR_o, B1-stability is now ruled out,
since a relatively large number of shells contributes significantly to the *three-ion energy* in this structure; these forces are
found to be repulsive.

The search for the origin of *selective* stability of B3, B4

is fascinating on the basis of the three-ion exchange model. We
present, in the following Table II, the results for three-ion
exchange energies, in units of nearest-neighbor (cation-anion)
repulsion, for BN, BAs, BeO and AlN in the B3- and B4-configu-
rations. The shells of anions around a central anion are numbered
by n̲.

Table II

Three-anion exchange energy in units of nearest-neighbor repulsion,
and without weighting factors, for BN, BAs, BeO and AlN in the
B3- and B4-structures.

		anion shell 1	2	3	4	5	6
BN	B3	-155	-15	-191	- 4	-30	-
	B4	-155	-15	+ 0.8	-123	-39.	-0.7
BAs	B3	-321	-33	-423	- 10	-75	-
	B4	-321	-33	+ 2	-270	-88	-1.5
BeO	B3	- 25	- 1	- 14	-	-	-
	B4	- 25	- 1	-	- 9	- 2	-
AlN	B3	- 16	- 1	- 12	-	-	-
	B4	- 16	- 1	-	- 8	- 2	-

The third shell of anions in B3 lies at the *same* distance
from the central anion as the fourth shell in B4. The numbers of
the table must still be multiplied by a *weighting factor* which
is positive if an anion of the corresponding shell and the central
anion repel each other, negative if they attract each other. We
verify at once that, if this factor is > 0 for the third shell in
B3 (and, therewith, also for the fourth shell in B4), then the B3-
structure has a more negative three-ion exchange energy *and is
thus more stable* . On the other hand, if this factor is negative,
then B4 will be the stable structure (the differences between
pair-interactions for each compound in B3 and B4 can be totally
neglected). The first possibility will be realized for smaller
values of the dimensionless parameter βR_o than the second case;
this is indeed confirmed by experiment (βR_o = 0.78 for BN and
0.75 for BAs, whereas the values are 0.99 for BeO and 0.93 for
AlN). Thus, in addition to accounting for B3, B4-stability relative to B1 and B2, the three-ion exchange forces stabilize B3 or
B4 *selectively*, as a function of βR_o and γ (at high values of γ).

C. *Stability of Ionic Solids of Composition* AX_2 [25]: Representatives are solid CaF_2, SiO_2, OCs_2, etc.; A is an element of columns II, IV or VI of the periodic table, X an element of columns I, VI or VII. The ideal crystals exhibit $(8,4)$, $(6,3)$ or $(4,2)$ coordination. The only lattice-type with $(8,4)$ coordination is fluorite (C1): to the second category $(6,3)$ belong rutile (C4), anatase (C5), cadmium iodide (C6) and cadmium chloride (C19). Structures with $(4,2)$ coordination are cuprite (C3), quartz (C8) and cristobalite (C9).

Again, attempts to explain the occurrence of all these structures, and to determine their relative stability, have been based on the HBM-model and on the more empirical Goldschmidt-model; for a review and summary of these analyses, see ref. [25]. It appears that, although the differentiation in stability is considerably more pronounced on the basis of the Goldschmidt-model (again due to the explicit relative-size parameter r_-/r_+ !) than in the framework of the HBM-description, essential structure-sensitive components are missing in the two analyses.

We have reconsidered this problem in the framework of the three-ion exchange model. Of the many interesting results of the analysis, we mention here only three aspects. Firstly, as was to be expected, the layer structures C6 (CdI_2) and C19 $(CdCl_2)$ are stable only if dipole-polarisation energy is taken into account. In addition, just as with B3, B4-*relative* stability, the three-ion exchange forces can distinguish between these two structures, again on the basis of three-ion contributions to the crystal energy from a *specific* shell of anions around a central anion (the third shell in C19, at the same distance from the central anion as the fourth shell in C6; see Table XV of ref. [25].). These considerations apply to e.g. $ZrSe_2$, stable in C6, with a βR_o-value of 1.08. For $MgCl_2$ and OCs_2, with βR_o-values of 1.42 and 1.47, respectively, those shells are too far away from the central ion to contribute significantly to the three-ion energy. The structures C6 and C19 then have practically the same crystal energy. Distortion of the C19-lattice, resulting in a (two-percent) increase of the Madelung energy [31], decides in favor of C19-stability.

Secondly, interesting results are obtained with respect to the relative stability of the β-quartz (C8) and β-cristobalite (C9) structures for SiO_2. These configurations occur simultaneously in nature and must, therefore, have practically the same

crystal energy. This, however, is not easily understood at first
sight: in cristobalite we find nearest-neigbor configurations
Si-O-Si with opening angle θ = 180° at the oxygen ion, whereas
in quartz θ < 180°. On electrostatic grounds one should then
expect cristobalite to be more stable than quartz. For this
reason, it is assumed that the quartz structure gains energy
through covalent Si-O bonding [32], entailing an opening angle
θ < 180°. Analysis on the basis of three-ion forces shows, first
of all, that the Si^{4+}-ion is so small that it may be replaced
by a *point charge*. Secondly, that cristobalite, although it has
a higher Madelung energy, loses this advantage because of strongly
repulsive exchange interactions between three *oxygen* ions, which
repulsion is practically absent in the quartz structure. This
result shows that we have here to do with repulsive three-ion
O^{2-}-respulsions and *not* with covalent Si-O bonding.

A third result which we may mention here concerns solid TiO_2
(βR_o = 1.15; γ = 4.2), which occurs both in a rutile-type (distorted C4-) and an anatase (C5-) modification. On the basis of
three-ion exchange forces superimposed on the HBM-model, it is
conclusively found that the stable structure is *fluorite* (C1),
not C4 or C5. The calculations are based on the assumption that
the Ti^{4+}-ion has the argon closed-shell configuration, i.e. that
the two 3d- and the two 4s-electrons have been split off. We are
thus led to conclude that this assumption is incorrect. The observed wide variety of compounds with different Ti-valence also
indicates the "flexibility" of Ti in adapting its electron configuration to the chemical environment.

D. *Polymorphism of Solids of Compounds with Closed d-Shell
Cations* [26]: Representatives are the halides of monovalent
copper and silver, as well as the chalcogenides of zinc and
cadmium. These solids are of particular interest because it is
found that their crystal energies are systematically more negative
than the values obtained on the basis of the HBM-model, with
relatively large discrepancies for the salts crystallizing in the
B3- or B4-structure (tetrahedral coordination). To account for
these deviations, Mayer [33] suggested already in 1933 that the
discrepancies are due to covalent bonding between the ions, but a
quantitative interpretation of this suggestion has never been
given. It is also remarkable that many of these salts exhibit
polymorphism in that they occur in a B3- as well as in a B4-

modification. This is in contrast with the B3-, B4-stability of
II-VI and III-V compounds (discussed above), where the cations
have rare-gas electron configurations. There, no such polymorphism is observed. The B3-, B4-stability for many of these
salts is in itself already abnormal, since the γ-values for the
compounds are not very high, contrary to the case of II-VI and
III-V compounds (CuF: γ = 1.37; ZnO : γ = 1.78; CdS : γ = 1.82,
but MgTe : γ = 15; AlN : γ = 16 as lowest values with B4-stability
for II-VI and III-V compounds. The lowest γ-value for B3-
stability is 26, for AlP).

An analysis [26] on the basis of the HBM-model plus three-
ion exchange forces gives a consistent explanation of the experimental data if it is assumed that short-range *attractive* cation-
cation (and possibly cation-anion) *pair interactions* are also
present. Such forces may arise, in second order of perturbation
theory, because the excited cation states lie relatively low
compared to those of cations isoelectronic with rare-gas atoms.
On this basis, the relative stability of the B1-, B3- and B4-
structures as well as the B3-, B4-polymorphism for many of these
salts can be explained, without introducing covalent bonding.
Furthermore, the model predicts correctly the sign, and order of
magnitude, of the deviations from the Cauchy relations between
elastic constants, as discussed below.

E. *Deviations from the Cauchy Relations between Elastic Constants*
[26,27]: Thus far, we have dealt with applications of the three-
ion exchange forces to problems of crystal stability for different
types of ionic solids. Such analyses suffer from the disadvantage
that crystal stability is determined by two-ion and three-ion
forces *together*, so that the information regarding three-ion
interactions is only indirect. However, as was mentioned in the
Introduction, measurements, at very low temperatures, of deviations
from the Cauchy relations between elastic constants in simple
ionic solids are determined *solely* by many-ion forces, if certain
conditions are satisfied. These conditions hold a.o. for the
structures B1 and B2. Here, we have only three independent elastic
constants c_{11}, $c_{13}(=c_{12})$ and $c_{44}(=c_{66})$. If the interionic potential
is pairwise additive, and if zero-point and thermal energies may
be neglected, then it is easily demonstrated that $c_{13} = c_{44}$ (a
Cauchy relation). Recently, measurements at liquid-helium temperatures have become available for many alkali halides (see ref. [27]
for details); they show that generally $c_{13} \neq c_{44}$. The experimental

values for $c_{13} - c_{44}$ are of either sign, although those with
negative sign are more numerous. We have already mentioned Löwdin's
analysis of these deviations [13]; later analyses are reviewed
in ref. [27]. For completeness, we mention that the general
expression (neglecting zero-point vibrations and thermal energy)
for the elastic constants is given by

$$c_{ij} = \frac{1}{V} \frac{\partial^2 E_{st}}{\partial e_i \partial e_j} \quad ; \quad i, j = 1, \ldots 6 \quad , \tag{4}$$

where V is the molar volume of the crystal and where the e_k
($k = 1, 2, \ldots, 6$) are the six Lagrangian components of the strain
tensor; E_{st} denotes the static lattice energy. The derivatives
are understood to apply for the initial stress-free state. If we
denote the many-ion crystal energy by ΔE_{st}, then we obtain,
from (4), the following expression

$$c_{13} - c_{44} = \frac{1}{V} \left(\frac{\partial^2}{\partial e_1 \partial e_3} - \frac{\partial^2}{\partial e_4^2} \right) \Delta E_{st} \quad . \tag{5}$$

In applying three-ion forces to the evaluation of $c_{13} - c_{44}$,
we find the following very simple general rules:

a) $\gamma \approx 1$ (*cation and anion of similar size*). The contribution to
$c_{13} - c_{44}$ by three-ion forces with two ions from a given shell
is *negative* if an ion from that shell and the central ion
repel each other; it is *positive* if this pair interaction is
attractive.

b) $\gamma \gg 1$ (*cation much smaller than anion*). The same rule as under
a) applies, now with regard to the anion shells around a
central anion. Three-ion forces involving one or more cations
give a negligible contribution to $c_{13} - c_{44}$.

On the basis of these rules, the observed deviations from
$c_{13} - c_{44}$ can be quantitatively interpreted in terms of contributions from the different shells around a central ion. In LiF,
for example, the first shell of Li-ions (weak) and the first shell
of F-ions (strong) around a central anion contribute to $c_{13} - c_{44}$;
their sum is strongly negative (about -2.25×10^{11} dyne/cm^2). In
LiI, on the other hand, the first shell of anions around the
central anion gives a negative, the second shell a positive contribution to $c_{13} - c_{44}$. The result is small and positive (about
$+0.05 \times 10^{11}$ dyne/cm^2). In the heavy Cs-salts we find a similar
compensating effect through the first and second shells of ions;
the result is again small (positive for CsCl and CsBr, negative

for CsI). In solid MgO, however, $c_{13} - c_{44}$ is strongly negative because of high electrostatic compression. The agreement with experiment is also quantitatively satisfactory. It should be noted that there is no direct correlation between the deviation $c_{13} - c_{44}$ and the relative B1-, B2-stability: for the heavy Cs-salts $c_{13} - c_{44}$ is small (positive or negative), but it is also small for NaBr (negative) and NaI (positive).

For salts of monovalent Cu and Ag, as well as of divalent Zn and Cd (closed 3d-shells), the (few) measured values for $c_{13} - c_{44}$ are without exception positive and rather large (AgCl: + 3.2 x 10^{11} dyne/cm^2; ZnS: +1.8 x 10^{11} dyne/cm^2; CdTe: + 2.3 x 10^{11} dyne/cm^2). According to the three-ion exchange model this implies that there exists an extra attraction between nearest- or nearneighbor ions in the lattice. We have above established the same fact in the analysis of crystal stability for such compounds. It is probable that one can obtain, on this basis, more insight into the cation-cation interactions suggested by van Arkel [34] in connection with the properties of TiO. The "abnormal" behavior of $c_{13} - c_{44}$ for salts with closed 3d-shells (large positive values) and their "abnormal" crystal structures (B3 and B4 occur frequently and as polymorphs) are thus seen to be directly correlated, and quantitatively understandable on the basis of three-ion exchange forces.

F. *Magnetic Ordering in Non-Conducting Solids with Paramagnetic 3d- or 4f-Cations* [28]: We will now briefly consider threeion exchange forces in non-conducting ionic solids of which the cations have an incomplete 3d- or 4f-shell (e.g. Mn^{2+}, Eu^{2+}, etc.). As mentioned in the Introduction, those are the systems in which three-ion forces were first studied by Kramers [1]. As we recall, Kramers used a three-center, four-electron model as the basic unit for superexchange, with one "effective" electron on each of two paramagnetic cations and two "effective", spin-paired, electrons on the diamagnetic anion. This system can be in a singlet or a triplet spin state, depending upon whether the spins of the electrons on the cations are antiparallel or parallel, respectively. Kramers' model for superexchange involved excited states of the basic unit in which an electron from the anion is excited to the orbit of the electron on the cation. He also supposed that the cation- and anion-orbitals may be taken as orthogonal to each other. As a consequence of these assumptions, indirect coupling

between the cation spins occurs first in *third* order of perturbation theory. For different analyses of the superexchange phenomenon, see Anderson [5].

In the exchange-perturbation theory, on the other hand, the
system of four electrons on three centers is considered *as a whole*,
and non-orthogonalities are rigorously taken into account [35]. As
a result, we find indirect coupling already in *first* order of the
perturbation analysis. Zeroth-order singlet- and triplet-wave-
functions can readily be constructed using projection-operator
techniques [36]. For the orbitals of the effective electrons we
take again simple-Gaussian functions, with characteristic para-
meters β (anion) and β' (cation), and with $\gamma = (\beta'/\beta)^2$ as a
relative-size parameter. The parameter β' for the cation refers
to the unpaired electrons only; it can be obtained from diamagnetic
susceptibilities, subtracting the part due to the closed electron
shells. For details of this procedure, we refer to [28a].

In zeroth order we assume complete ionicity, i.e. we start
from the systems $Ni^{2+}O^{2-}$, $Mn^{2+}S^{2-}$, etc. The quantity calculated
is the "effective" cation-cation interaction, defined by eq.(2)
for cations $\underline{a}$ and $\underline{c}$. As an illustration of the superexchange
effect we present, in the following Fig. 1, the first-order
effective cation-cation interaction for a configuration $Mn^{2+}-S^{2-}-$
Mn^{2+}, as a function of the angle θ at the sulfur anion, for the
triplet (T) and singlet (S) configurations. The dashed curves represent the direct exchange between the Mn^{2+}-ions, i.e. without
the S^{2-}-ion [28b].

From the curve we infer the essential importance of the diamagnetic anion for the (effective) interaction between the two
cations, especially for values of the opening angle $\theta > 60^\circ$.
Second-order results are found to be quite similar to those in
first order. In view of this similarity we expect that, in a first
approach, stability criteria for different types of magnetic
ordering can be established on the basis of first-order interactions alone.

The results can now be applied to the determination of the
most stable spin arrangement in solids, of composition CA, of
3d- and 4f-compounds. The dominant crystal structure is B1, with
four different possible antiferromagnetic arrangements of cation
spins (denoted by AF1 -AF4). In addition, of course, we have to
consider stability of a ferromagnetic pattern, with all spins
parallel. In view of the fact that we do not include spin-lattice

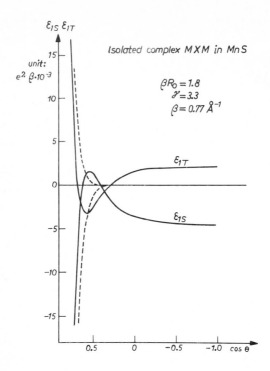

Fig. 1. Superexchange plus direct exchange interactions for an
isolated complex MXM of two paramagnetic cations M and one
diamagnetic anion X forming an isosceles triangle with
equal X-M distances, in first order of exchange perturbation theory, as a function of the opening angle θ at
the anion, as given by the three-center four-electron
model. Both the singlet (index S) and triplet (index T)
states are given. $\beta = 0.77$ Å^{-1} (S^{2-}), $\beta' = 1.4$ Å^{-1} (Mn^{2+}),
$\gamma = 3.3$ and $\beta R_0 = 1.8$. The dashed curves represent the
corresponding interactions in the absence of the diamagnetic anion X.

interactions in the Hamiltonian, it is not possible to determine
the direction of easy magnetization. The large majority of 3d-
compounds with the B1-structure exhibits at the lowest temperatures antiferromagnetic ordering of the second kind (AF2), in
which one-half (6) of the first neighbors have spins parallel to
that of the reference cation, the other half antiparallel. The
spins of all (6) second neighbors are antiparallel to that of the
reference cation. For 3d-compounds the only known exception is
CrN, which is antiferromagnetic of the fourth kind. All 4f-compounds
which are non-conductors crystallize in the B1-structure and most
of them also exhibit antiferromagnetic ordering of the second kind.

A few (EuO, EuS and GdN) are ferromagnetic, whereas EuSe and some
others show a more complicated behavior. The Néel temperatures
(antiferromagnetic to paramagnetic) are of the order of a few
hundred degrees Kelvin for the 3d-compounds, between a few and
30°K for the rare-earth compounds. The Curie temperatures (ferromagnetic to paramagnetic) are 70°K (EuO), 60°K (GdN) and 16°K (EuS).

For an illustration of the results obtained on the basis of
three-ion exchange forces, we select solid MnS [28b]. This compound is of particular importance since it crystallizes in three
different structures: B1(NaCl), B3(sphalerite) and B4(wurzite).
It is often supposed in the literature that directed, covalent,
cation-anion bonding plays an important rôle in phenomena of
magnetic ordering, as, for example, in the Goodenough-Kanamori
rules [37]. If this should be the case then we must expect that
the experimental data on MnS in its three different modifications
cannot be explained on the basis of a simple model of three-ion
exchange forces between effective electrons on spherically-
symmetric (Gaussian) orbitals. The observed differences (AF2 in
the B1-, AF3 in the B3-, (B4-)modification) should then be
ascribed to different covalent character of the Mn-S bonding in
an octahedral (B1) and a tetrahedral (B3, B4) environment.

In summing over all possible cation-anion-cation complexes
we can determine, starting from a reference anion, the magnetic
energies in solid MnS for each structure (B1, B3 and B4) and for
the four types of antiferromagnetic ordering (AF1-AF4), as well
as for a ferromagnetic pattern. It is convenient to analyze
magnetic stability in terms of transition temperatures from the
ordered to the disordered structures (Néel or Curie temperatures),
the pattern with the highest transition temperature being the most
stable one. These transition temperatures are, in a molecular-
field approximation, obtained as linear combinations of coupling
parameters J_i, defined in the framework of a Heisenberg effective-
spin Hamiltonian; J_1 refers to (effective) interaction between
nearest-neighbor cations, J_2 to that between second neighbors, etc.
For details we refer to [28]. The results of the three-ion exchange
model for MnS are given in the following table.

The stability of antiferromagnetic ordering of the second kind
(AF2) in the B1-modification is found to be due mainly to the
effect of strong $\theta = 180°$-superexchange coupling with antiparallel
spins. This type of coupling is completely absent in B3 (and B4);
for this reason, AF2 cannot be stable in those structures.

Table III

Results for coupling coefficients J_1 and J_2, magnetic crystal energy ε_{cr} and Néel temperatures, of solid MnS in its three modifications, obtained on the basis of the present model as compared with experiment.

Exp.	Structure					
	R_o (Å)	2.606	2.425		2.427	
	J_1 (°K)	-8.4	-12.4		-10.7	
	Magnetic ordering	second kind	third kind		third kind	
	T_N (°K)	154	~150		~100	
Calc.	J_1 (°K)	-7.0	-10.2		-10.1	
	J_2 (°K)	-4.2	-0.02		-0.005	
	Magnetic ordering	second kind	first kind	third kind	first kind	third kind
	T_N (°K)	144.0	237.0	237.5	236.1	236.3
	ε_{cr} (kcal/mole)	-1.744	-2.124	-2.126	-2.114	-2.115

In B3 and B4, antiferromagnetic ordering of the third kind (AF3) is indeed found to be the most stable one, but the difference with AF1 is clearly too small. The reason, we expect, is that a complete analysis should also incorporate two cation-two anion exchange (represented, in our model, by a four-center, six-electron model). The general good agreement with experiment indicates that differences in covalent Mn-S bonding between B1 and B3 (B4) can play only a minor rôle regarding magnetic ordering in solid MnS.

The observed *ferromagnetic* ordering in EuO, EuS and GdN presents a major problem to the theory [38]. The sharp drop in Curie-temperature between EuO and EuS (from 70°K to 16°K) and the antiferromagnetism observed with EuTe indicate that the origin of ferromagnetism must lie in *direct* cation-cation exchange; if superexchange were involved, then, upon replacing the O^{2-}-ion by S^{2-}, the indirect exchange would increase because of the larger anion, and decrease because of the larger cation-anion separation. On the other hand, direct cation-cation exchange decreases unilaterally because of larger separation. The search, consequently, is for the origin of ferromagnetic *direct* cation exchange. Direct interactions between electrons on 4f-orbitals of different cations

cannot be expected to contribute significantly to the magnetic
energy, since these orbitals are buried inside their respective
valence shells. Therefore, *some kind* of "indirect" exchange
between 4f-electrons on different cations must take place. We have
proposed [28a] that 4f-electrons on different cations can interact indirectly through *their own valence shells* (so-called "indirect valence-shell exchange"). In the present model, re-
presenting the valence shells of each cation by two spin-paired
effective electrons, the model is that of a two-center, six-
electron system (three electrons per cation). Preliminary calculations [28a] show indeed that, on this basis, EuO and EuS
should be ferromagnetically, the heavier Eu-chalcogenides anti-
ferromagnetically ordered, as a result of competition between
(antiferromagnetic) superexchange and (ferromagnetic) indirect
valence-shell exchange. For details, we refer to [28a].

Let us, at this point, take a moment's pause. We have considered ionic solids principally because electrostatic compression
brings the ions so close together that many-ion forces are likely
to become important. We have developed a shell-by-shell description of such many-ion effects which accounts for relative stability of solids of alkali halides, II-VI and III-V compounds, those
of composition AX_2, and those with closed-shell 3d-cations. We
have verified that the same model gives a consistent explanation
for the observed low-temperature deviations from the Cauchy
relation $c_{13} = c_{44}$ in B1 and B3 (B4) between elastic constants.
After that, we have established that the same type of model is
also applicable to magnetic ordering in solids of compounds with
incomplete 3d- or 4f-shells of cations. All considerations were
based on a HBM-model *plus* three-ion exchange forces. Covalent
(homopolar) forces were found to be non-essential in all examples
treated in detail (B3-, B4-relative stability, C8-C9 relative
stability for SiO_2; C6-C19 relative stability for $ZrSe_2$; magnetic
ordering in MnS). We can then, finally, ask the question whether
the model is also applicable to a variety of *atomic* systems,
since "exchange" *per se* has nothing to do with ionic character
of interacting particles. Indeed, as we recall, the first
application concerned rare-gas crystal stability [19], although
the three-atom exchange interactions in such solids are orders
of magnitude weaker than in ionic crystals.

G. *Stability of Rare-Gas Compounds* [29]: Of *open-shell* atomic
systems there exist a number of very interesting simple examples,
such as the fluorides of the rare-gas atom Xe and Kr (XeF_2, XeF_4
and XeF_6 as well as KrF_2 and KrF_4). These compounds are stable,
with a binding energy of about 15-30 kcal/mole bond; their discovery
in 1962 [39] came, consequently, as a great surprise. Many efforts
have since then been undertaken to explain stability, selectivity
and geometric conformations on the basis of different theoretical
approaches (hybridisation, resonance, spin correlations, and
molecular orbitals in localised and delocalised versions), with
varying success. The most conspicuous properties of these molecules are: (1) only xenon and krypton as rare-gas atoms and only
fluorine as halogen atom form stable compounds; (2) XeF_2 is a
linear molecule, XeF_4 is square-planar and *not* tetrahedral; (3)
xenon and krypton form compounds only with an *even* number of
fluorine atoms. It appears that, generally, the theoretical methods
possess too low a selectivity to explain all these properties, even
in a semi-empirical version [40].

   A fluorine atom has an incomplete outer p-shell and spin 1/2,
represented in the model by one effective electron and a nuclear
charge +1. A rare-gas atom with its closed shells is replaced by
two spin-paired effective electrons and a nuclear charge +2. A
complex XRX, with X a halogen atom, R a rare-gas atom, is then
represented by the *same* three-center, four-electron model as we
have already applied in the discussion of magnetic ordering in non-
conducting ionic solids. The extension to molecules RX_n, with
n > 2, is straightforward. We can now, again on the basis of Gauss-
functions with parameters β (rare-gas atom), β' (halogen atom) and
with $\gamma = (\beta'/\beta)^2$ as a measure of their relative size, simply take
over the results of the calculations for magnetic order. In the
following figure the results are given for the binding energy (sum
of first- and second-order perturbations) of linear rare-gas
dihalides as a function of γ. Values of γ < 1 imply that the
halogen atom is "larger" than the rare-gas atom; for γ > 1 the
situation is reversed.

   For XeF_2, γ is about 2.5 and the model predicts this molecule
to be stable. For KrF_2, γ ≈ 1.8; also this molecule is predicted to
be stable, although more weakly bonded than XeF_2. If we consider
$XeCl_2$, then γ decreases to approximately 0.85. From the curve we
conclude at once that this molecule cannot possibly be stable;

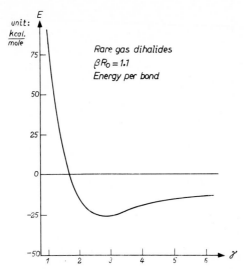

Fig. 2. Sum of first- and second order interactions for linear
rare-gas dihalides as a function of the relative size
of the rare-gas and halogen atoms. Binding energy is
given in kcal/mole and per bond. For Xe,F γ is 2.5, for
Xe,Cl: 0.85, for Kr,F: 1.8 and for Kr,Cl: 0.6.

its existence has not been observed. Regarding geometric conformation, the model is very selective: XeF_2 with an opening
angle θ = 90° at the xenon-atom is found to exhibit no binding
at all.

With molecules RX_n and n > 2 we find very analogous results.
First of all, no stability is obtained for odd values of n; this
seems to be related to the fact that for odd n a total-spin state
S = 0 is not possible. States with S ≠ 0 are always unstable in
the model; this observation applies as well for even values of n.
As an illustration, we show, in the following two figures, the
binding energy of XeF_4 in different spin states [29b], first with
a square-planar configuration (Fig. 3) and then with tetrahedral
shape (the value of βR_c lies between 1.1 and 1.2; for the
method of determining β, β'-values, see ref. [29a]).

From the figures we see that only a *square-planar* XeF_4 with
S = 0 is stable. It is important to note that there is no question
of "hybridisation" of orbitals on the Xe-atom upon molecule
formation: the molecule can, in good approximation [29b], be described in terms of "superexchange building blocks" XeF_2, in con-
flict with the concept of directed valence.

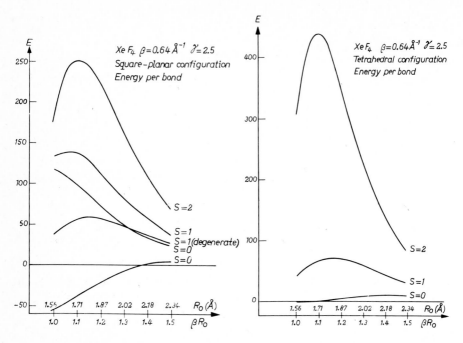

Fig. 3/4. Calculated energy per bond, in kcal/mole, for the molecule XeF$_4$ as a function of the dimensionless parameter
βR_o, where R_o is the Xe-F distance.

H. *Rotational Barriers in Simple Organic and Inorganic Molecules*
[30]: Finally, we will briefly outline a recent application of the
model to the interpretation of the rotational barriers in simple
molecules, limiting ourselves to rotations about *single* bonds such
as in CH_3-CH_3, BCl_2-BCl_2, $SiCl_3$-$SiCl_3$, etc. The existence of such
barriers was established about forty years ago (ethane); since
then, they have been the subject of numerous experimental and
theoretical studies. Excellent reviews on the subject have been
written, e.g. by Wilson [41] on its early history and by Lowe [42]
and Bastiansen, Seip and Boggs [43] on recent developments in the
field.

        Early theories were largely empirical in character. Molecules
such as C_2H_6, C_2F_6 and other fully-halogenated ethane derivatives
are observed to be stable in the so-called "staggered" configuration, in which the atoms at the two ends of the molecule are
turned away as much as possible with respect to each other. The
barrier is given by the difference in energy between the "eclipsed"
conformation (atoms at different ends are on top of each other in

projection) and that of the staggered configuration. In BCl_2-BCl_2, the dihedral angle (between the two BCl_2-planes) is $90°$. It is, therefore, intuitively tempting to ascribe the origin of the barrier to repulsive interactions between non-bonded atoms at different ends of the molecule. This repulsion can be simulated by means of ad hoc pair-potentials with adjustable parameters.

Quantitative theory entered the stage with the pioneering SCF-LCAO study of Pitzer and Lipscomb [44] on the rotational barrier in ethane. The calculated value of the barrier was 3.3 kcal/mole, compared with an experimental barrier of 2.88 kcal/mole. However, the calculated energies of the staggered as well as the eclipsed configuration separately were found to be too high by about 160 times the barrier height. Apparently, practically complete cancellation of errors had occurred in determining the difference. Important further progress has recently been made by Kern, Pitzer and collaborators [45] on the basis of "bond-orbital" calculations for ethane and methanol. The wavefunction for a CH- (or OH-) bond was written as a linear combination of a hybridized C- (or O-) orbital and a 1s-orbital on hydrogen. A parameter λ measures the "polarity" of the bond. The most essential result of the analysis was that reasonable values for the barrier are obtained, over a *wide range* of λ, if the total wavefunction is *constrained to satisfy the Pauli principle* . This means that the barrier originates from exchange interactions between bonds, and that the precise description of the wavefunction for each bond is not of relevance.

We have applied the model of many-atom exchange forces to this problem [30], ignoring the atoms at the ends of the axis of rotation, and replacing the electrons on each of the other atoms by one effective electron, with spins parallel. For C_2H_6, Si_2Cl_6, etc. this leads to a six-center, six-electron model, whereas for B_2Cl_4 there are four effective electrons on four centers. Once more, we adopt simple-Gaussian functions for the electron orbitals and use the same β'-parameters as determined in the analysis of rare-gas halides. The only unknown is β_H; it is chosen such that the ethane barrier is reproduced. The zeroth-order wavefunction for the system is of the form $(n < 6)$

$$\psi^{(o)} = A_n \chi_n, \tag{6}$$

where A_n is the antisymmetrizer for the system considered and χ_n is the simple product of the $\underline{n}$ wavefunctions for the effective electrons separately ($\psi^{(o)}$ is a basis vector for the alternating representation of the permutation group of $\underline{n}$ elements). Symmetry

considerations substantially simplify the expression for A_n
(which otherwise would contain n! permutations). For details, we
refer to [30].

    We have carried out the calculations, in a *rigid*-rotor
approximation, in *first* order of perturbation theory and for one
molecule (B_2Cl_4) also in *second* order. In the latter case it is
found that the interaction energies, as a function of the angle
of rotation θ, are very similar in the two orders. Since the distances between the atoms in the molecules are relatively small,
we assume again that the average excitation energy E_{Av} is large
and, thus, that second-order forces are generally suppressed. The
only exceptions are molecules in which a hydrogen atom is bonded
to a silicon or a germanium atom; the Si-Si distance e.g. is
2.32 Å, the C-C distance only 1.54 Å. In such cases, it must be
expected that E_{Av} is lower than for molecules with a C-C axis
of rotation. Thus, second order may also contribute significantly
to the barrier. In accordance with this prediction, we find
first-order barriers which are consistently too low for molecules
with hydrogen attached to Si or Ge on the axis of rotation.

    In the following figure the values for the barrier of rota-
tion are presented, as a function of the Gaussian parameter β
of the non-bonded atoms, for the molecules C_2H_6, C_2Cl_6, B_2Cl_4
and Si_2Cl_6.

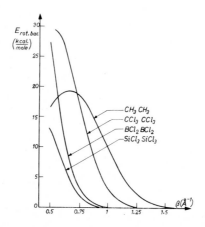

<u>Fig. 5</u>. First-order rotational barriers $E_{rot.bar.}$ for the mole-
cules C_2H_6, C_2Cl_6, B_2Cl_4 and Si_2Cl_6, as a function of the
Gaussian parameter β of the hydrogen or chlorine atom.

With β_{Cl} = 0.8 $\overset{\circ}{A}{}^{-1}$, as used for rare-gas compounds, the calculated barrier heights are found to be 13.4 kcal/mole for C_2Cl_6 (exp. between 10.8 and 17.5 kcal/mole), 1.7 kcal/mole for B_2Cl_4 (exp. 1.5 $\pm$ 0.6 kcal/mole) and 1.05 kcal/mole for Si_2Cl_6 (exp. ~ 1 kcal/mole). For all these molecules, the stable conformation is staggered, in agreement with that observed. The experimental barrier for ethane (2.88 kcal/mole) corresponds with β_H = 1.22 $\overset{\circ}{A}{}^{-1}$; we have taken a value of 1.2 $\overset{\circ}{A}{}^{-1}$ in all calculations. Results for the barrier heights of approximately forty molecules are found to be in good agreement with experiment, with the few exceptions mentioned earlier [30]. The model was also applied to substituted ethanes of the type H_2XCCYH_2, where X and Y denote halogen atoms. Two conformers, *gauche* and *trans*, are observed for these molecules. The predicted stable conformer is always the one observed, with reasonable agreement for the energy difference. For details, we refer to [30].

We have undertaken a *cluster decomposition* of the first-order energy, as a function of the angle of rotation θ. The first term (E_2) of the series represents the contribution by *pair* interactions to the energy, the second term (ΔE_3) the contribution by *three-atom* interactions, then ΔE_4, ΔE_5 and ΔE_6, referring to simultaneous interactions between four, five and six atoms, respectively. This decomposition, for the ethane molecule, is presented in the following figure (θ = 0° corresponds with the eclipsed, θ = 60° with the staggered conformation. Note the different scales).

Two features of this decomposition are of particular interest:

(a) the many-body contributions to the interaction energy for both the eclipsed and staggered conformations *alternate in sign* with increasing number of constituent atoms involved;

(b) the contribution to the barrier height is *positive* for two-, three- and four-atom interactions, negative for those between five atoms and very small (positive) for simultaneous interactions between all six atoms. A rough approximation to the height of the barrier can already be obtained by considering only ΔE_2 and ΔE_3, since ΔE_4 and ΔE_5 are of opposite sign and have roughly the same value. The cluster expansion for a *given* conformation converges quite rapidly with the number of interacting atoms.

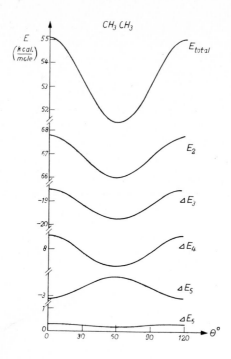

<u>**Fig. 6.**</u> Contributions to interaction energies (in kcal/mole) in
      CH_3CH_3, according to the number of interacting substituents,
      as a function of the dihedral angle θ .

   It is found that very similar results hold for C_2Cl_6 so that,
probably, the above characteristics are of general validity [46].

   As a last remark, we note that the present analysis is clearly not applicable to C-C *double* bonds as axes of rotation, such
as in ethylene. This molecule is *planar* and has a barrier height
of 65 kcal/mole, very much higher than the barriers we have discussed here. Apparently, the axis of rotation plays an important
rôle in determining the barrier. There may then exist a correspondence with the stability of the *xenon tetrafluoride* molecule,
which, as we know, is square-planar and which, as we have found,
is stabilized by *superexchange* interactions. Research on this
subject is in progress.

References

[1] H.A. Kramers, Physica 1 (1934) 182.

[2] J. Becquerel, W.J. de Haas and J. van den Handel, Physica 1 (1934) 383.

[3] D. de Klerk, Physica 12 (1946) 513.

[4] C.G. Shull and J.S. Smart, Phys. Rev. 76 (1949) 1256.

[5] P.W. Anderson, in: Solid-State Physics, Vol. 14, Eds. F. Seitz and D. Turnbull (Academic Press, New York, 1963) p.99.

[6] C. Herring, in: Magnetism, Vol. IIB, Eds. G.T. Rado and H. Suhl (Academic Press, New York, 1968) p. 1.

[7] S.V. Vonsovsky and B.V. Karpenko,in: Encyclopedia of Physics, Vol. XVIII/1, Ed. S. Flügge (Springer, Berlin, 1968) p. 265; S. Methfessel and D.C. Mattis, ibid, Vol. XVIII/1, p. 389.

[8] B.M. Axilrod and E. Teller, J. Chem. Phys, 11 (1943) 299.

[9] B.M. Axilrod, J. Chem. Phys. 19 (1951) 724.

[10] J.S. Brown, Phys. Letters 29A (1969) 121, and references therein.

[11] R. Fowler and H.W. Graben, J. Chem. Phys. 56 (1972) 1917.

[12] H. Margenau, Rev. Mod. Phys. 11 (1939) 1.

[13] P.O. Löwdin, A Theoretical Investigation into Some Properties of Ionic Crystals (Almqvist and Wiksell, Uppsala, 1948).

[14] P. Rosen, J. Chem. Phys. 21 (1953) 1007.

[15] A. Shostak, J. Chem. Phys. 23 (1955) 1808.

[16] B.G. Dick and A.W. Overhauser, Phys. Rev. 112 (1958) 90.

[17] L. Jansen, Phys. Rev. 162 (1967) 73; L. Jansen and E. Lombardi, Chem. Phys. Letters 1 (1967) 417.

[18] W. Byers Brown, Chem. Phys. Letters 2 (1968) 105; S. Farberov, V.Ya. Mitrofanov and A.N. Men, Intern. J. Quantum Chem. 6 (1972) 1057.

[19] R.T. McGinnies and L. Jansen, Phys. Rev. 101 (1956) 1301; L. Jansen and R.T. McGinnies, Phys. Rev. 104 (1956) 961.

[20] A. Dalgarno and N. Lynn, Proc. Phys. Soc. (London) A69 (1956) 821.

[21] A. van der Avoird, Chem. Phys. Letters 1 (1967) 429.

[22] L. Jansen, Phys. Rev. 135 (1964) A1292.

[23] E. Lombardi and L. Jansen, Phys. Rev. 136 (1964) A1011; for a review, see L. Jansen, Mod. Quantum Chem. 2 (1965) 239.

[24] E. Lombardi and L. Jansen, Phys. Rev. 140 (1965) A275;
for a review, see L. Jansen and E. Lombardi, Disc. Faraday
Soc. 40 (1965) 78.

[25] E. Lombardi and L. Jansen, Phys. Rev. 151 (1966) 694.

[26] E. Lombardi and L. Jansen, Phys. Rev. 185 (1969) 1158.

[27] E. Lombardi, L. Jansen and R. Ritter, Phys. Rev. 185 (1969)
1151, 1158;

E. Lombardi and R. Ritter, Chem. Phys. Letters 7 (1970) 143.

[28] (a) R. Ritter, L. Jansen and E. Lombardi, Phys. Rev. B8
(1973) 2139;

(b) L. Jansen, R. Ritter and E. Lombardi, Physica 71 (1974)
425.

[29] (a) E. Lombardi, R. Ritter and L. Jansen, Intern. J. Quantum
Chem. 7 (1973) 155;

(b) E. Lombardi, L. Pirola, G. Tarantini, L. Jansen and
R. Ritter, ibid 8 (1974) 335.

[30] E. Lombardi, G. Tarantini, L. Pirola, L. Jansen and R. Ritter,
J. Chem. Phys. 61 (1974) 894.

[31] Q.C. Johnson and D.H. Templeton, J. Chem. Phys. 34 (1961) 2004.

[32] See, e.g., R.C. Evans, An Introduction to Crystal Chemistry
(Cambridge Univ. Press, New York, 1952) Ch. VII.

[33] J.E. Mayer, J. Chem. Phys. 1 (1933) 327.

[34] See, e.g., A.E. van Arkel, Moleculen en Kristallen (Van
Stockum, 's-Gravenhage, 1961) Ch. XIII.

[35] A critical discussion of the assumptions underlying Kramers'
model as well as other superexchange mechanisms has been
given by R. Block, Ph.D. Thesis, University of Amsterdam
(1974) Ch. I.

[36] For a review of this method, see ref. [28a].

[37] See, e.g., P.W. Anderson, ref. [5];

J.B. Goodenough, Magnetism and the Chemical Bond (Inter-
science, Publishers New York, 1963).

[38] For an excellent discussion, see S. Methfessel and D.C.
Mattis, ref. [7].

[39] H.H. Claasen, H. Selig and J.G. Malm, J. Am. Chem. Soc. 84
(1962) 3593;

R. Hoppe, W. Dähne, H. Mattauch and K.M. Rödder, Angew.
Chem. 74 (1962) 903.

[40] C.A. Coulson, J. Chem. Soc. A (1964) 1442;

J.G. Malm, H. Selig, J. Jortner and S.A. Rice, Chem. Rev.
65 (1965) 199;

H. Selig, in: Halogen Chemistry, Vol. I, Ed. V. Gutmann (Academic Press, New York, 1967).

[41] E.B. Wilson, Jr., Advan. Chem. Phys. 2 (1959) 367.

[42] J.P. Lowe, Proc. Phys. Org. Chem. 6 (1968) 1; Science 179 (1973) 527.

[43] O. Bastiansen, H.M. Seip and J.E. Boggs, in: Perspectives in Structural Chemistry, Vol. IV, Eds. J.D. Dunitz and J.A. Ibers (Wiley, New York, 1971) p. 60.

[44] R.M. Pitzer and W.N. Lipscomb, J. Chem. Phys. 39 (1963) 1995.

[45] O.J. Sovers, C.W. Kern, R.M. Pitzer and M. Karplus, J. Chem. Phys. 49 (1968) 2592; C.W. Kern, R.M. Pitzer and O.J. Sovers, J. Chem. Phys. 60 (1974) 3583.

[46] Recently, G.F. Musso and V. Magnasco, Chem. Phys. Letters 23 (1973) 79, have also undertaken an analysis of the rotational barrier in ethane in terms of a cluster expansion, on the basis of a bond-orbital approach. Their results are (in part) very similar to the ones reported here.

AUTHOR INDEX

239

244